D0944624

Dispersion Forces

COLLOID SCIENCE

Editors
R. H. Ottewill and R. L. Rowell

In recent years colloid science has developed rapidly and now frequently involves both sophisticated mathematical theories and advanced experimental techniques. However, many of the applications in this field require simple ideas and simple measurements. The breadth and the interdisciplinary nature of the subject has made it virtually impossible for a single individual to distill the subject for all to understand. The need for understanding suggests that the approach to an interdisciplinary subject should be through the perspectives gained by individuals.

The series consists of separate monographs, each written by a single author or by collaborating authors. It is the aim that each book will be written at a research level but will be readable by the average graduate student in chemistry, chemical engineering or physics. Theory, experiment and methodology where necessary are arranged to stress clarity so that the reader may gain in understanding, insight and predictive capability. It is hoped that this approach will also make the volumes useful for non-specialists and advanced undergrauates.

The author's role is regarded as paramount and emphasis is placed on individual interpretation rather than on collecting together specialist articles.

The editors simply regard themselves as initiators and catalysts.

1. J. Mahanty and B. W. Ninham: Dispersion Forces

Dispersion Forces

J. MAHANTY
Department of Theoretical Physics

B. W. NINHAM
Department of Applied Mathematics

Research School of Physical Sciences
Australian National University
Canberra, A.C.T.
Australia

1976

ACADEMIC PRESS

LONDON · NEW YORK · SAN FRANCISCO
A Subsidiary of Harcourt Brace Jovanovich, Publishers

ACADEMIC PRESS INC. (LONDON) LTD.
24/28 Oval Road,
London NW1

United States Edition published by
ACADEMIC PRESS INC.
111 Fifth Avenue
New York, New York 10003

Library of Congress Catalog Card Number: 75-27236
ISBN: 0-12-465050-3

Filmset by The Universities Press, Belfast

Printed in Great Britain by
Whitstable Litho, Whitstable, Kent

Preface

This book was written by two mathematical physicists and deals with a subject which has been traditionally the province of colloid chemists. As such, it is bound to be a mixed bag. Our main reasons for accepting a bridging commission to write about the theory of dispersion forces and applications in some of the "grey areas" that join chemistry and physics, microscopic and macroscopic phenomena are that (i) we felt that a simple semi-classical formulation of the theory of the sort developed here would be useful in unifying a field which has been the subject of an extensive and complicated literature of late, and (ii) through comparing theory with experiment in a variety of situations which arise in the "grey area", people interested in the calculation of actual force fields would be able to work out and apply simple recipes based on the approach presented here. The book is intended for the use of graduate students and researchers in areas spanning physics, chemistry and biology where dispersion forces play a role.

The "grey area" is vast in scope, encompassing colloid science, and more specifically, surface physics and chemistry, adsorption, polymers, electrolytes, solutions generally, and impinges on a large chunk of modern biology. It is neither our intention, nor within our competence to cover comprehensively every facet of these fields. We expect specialists to be somewhat dissatisfied. For example, liquid state physicists will protest that we have hardly mentioned the role of structure due to short-range repulsive forces; others that we have not gone further to deal with hydrophobic or hydrophilic solutes; adsorption people will be aggrieved that we have not elucidated the limits of validity of the B.E.T. theory in practical solutions. Nonetheless we will be happy if this book stimulates interest in the field by revealing the similarity of concepts and techniques used here and in other branches of physics and chemistry, particularly for the uninitiated working in other disciplines; as it was for us.

One of us (B. W. N.) was introduced to the subject of this book by V. Adrian Parsegian, and owes much to a stimulating collaboration spanning several years. We have benefited much from knowing and working with Dieter Langbein. We have enjoyed working with past and present colleagues: E. Barouch, C. J. Barnes, D. Y. C. Chan, B. Davies, J. N. Israelachvili, D. J. Mitchell, J. W. Perram, D. D. Richardson, P. Richmond, K. W. Sarkies, R. A. Sammut, E. R. Smith and L. R. White, and during the process drank too much beer at the ANU Staff Centre on Lake Burley Griffin, which made the whole business worthwhile. A. G. de-Rocco, G. H. Weiss, R. H. Ottewill, S. Levine and D. H. Everett gave us much encouragement.

Norma Chin astonished us by correcting most of our errors, typing a complete manuscript in three weeks without any fuss or bother whatever. The manuscript could not have been completed without the meticulous care and interest with which she saw the book through to completion. We are in her debt.

The term "dispersion forces" is used throughout this book as a synonym for van der Waals forces, that is, those due to electrodynamic fluctuations. Those primarily of electrostatic origin or (directly or indirectly) due to electron overlap, or chemical bonds are not considered.

Contents

Chapter 3   Calculations and Comparison of Theory with Experiment

Chapter 4   Dispersion Forces and Molecular Size

Chapter 5   Geometry and Anisotropy

To
Norma, Jenny and Sarojini
for their patience

Chapter 1

Introduction

1.1 Force between molecules

The origin and nature of intermolecular forces have been analysed and studied since the development of the atomic hypothesis which led to the picture of bulk matter having a microscopic structure in terms of the constituent molecules. The application of such a picture to explain experimental data on gases, perhaps more than anything else, revealed the need for greater understanding of intermolecular forces. In the early days of kinetic theory, molecules were regarded as hard objects with negligible long range interaction between them. This model was adequate to explain some of the simpler properties of gases, Boyle's and Charles' laws, and viscosity and thermal conductivity. But properties like the detailed features of the equation of state of a gas could not be explained in terms of such a model. It was recognized that a more realistic model would involve the concept of a force field between any pair of molecules whose range is larger than molecular dimensions.

One of the most successful applications of the concept of the inter-molecular force field is to the equation of state of a gas given by van der Waals (1873). In view of the importance of this equation, we shall briefly consider the way in which the intermolecular force field occurs in it. Let us consider a gas with n molecules per cubic centimetre, enclosed in a box of volume V. We shall assume that there is no interaction between the wall and the gas molecules. In effect, the available volume would be

$(V-b)$, where b is related to the size of all the molecules, so that $n = N/(V-b)$, N being the total number of molecules. When a molecule is at A, i.e., in the interior (see Fig. 1.1), its kinetic energy is $\frac{3}{2}k_B T$ where k_B is Boltzmann's constant. But when it is at B, i.e., at the wall of the container, its kinetic energy will be somewhat less, because in coming to the surface it would have done some work against the attractive forces exerted by the other molecules. Thus, the pressure exerted by the molecules colliding against this wall will be somewhat less than in an ideal gas, and this difference would depend on the density. Van der Waals suggested that this density dependence would arise from two sources— firstly, the number striking unit area of the wall per unit time which is proportional to the density, and secondly, the intensity of the field of force due to the other molecules (against which work is done by the molecule coming to the surface). This may be assumed to be proportional to the density if the forces are of a range larger than molecular dimensions, but smaller than the size of the vessel. Thus, the observed pressure has to be corrected by adding a term Kn^2 which is proportional to the square of the density to give the equivalent ideal gas pressure, i.e., $(P + Kn^2) = nk_B T$. Since $b \ll V$, this equation can be written in the form given by van der Waals,

$$\left(P + \frac{a}{V^2}\right)(V-b) = Nk_B T. \tag{1.1}$$

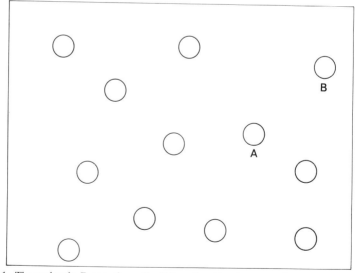

FIG. 1.1. The molecule B near the wall will have less kinetic energy than molecule A in the interior, because of attraction by other molecules. Cohesive forces between the wall and B are ignored.

These arguments, although not rigorous, do bring out the important point that the constant "a" occurring in the van der Waals equation of state is directly related to the strength of intermolecular forces. Subsequent developments in gas theory using statistical mechanics have established relationships between the virial coefficients in the virial expansion of the equation of state and the strength and form of intermolecular forces. The second virial coefficient obtained from the van der Waals equation of state is,

$$B = b - (a/RT), \tag{1.2a}$$

where R is the gas constant, and from statistical mechanics this is,

$$B = \int [e^{-V(r)/k_B T} - 1] \, d^3 r. \tag{1.2b}$$

Despite the qualitative character of the arguments used by van der Waals in his derivation, the great success of his equation of state in explaining experimental data on properties of gases spurred thinking on the origin of molecular forces.

The initial efforts in this direction were mainly empirical in character. Historical description of the earlier phases of the development of the subject of intermolecular forces is given by Hirschfelder, Curtiss and Bird (1954), and by Margenau and Kestner (1971), and will not be repeated here in detail. We shall mention only those theories which aimed at explaining molecular forces in terms of atomic structure. One of the first of several such attempts was by Reinganum (1912) (see also J. J. Thomson, 1914), who developed a dipolar model of the molecule, and obtained an interaction energy proportional to $1/R^6$ by averaging the interaction between two dipoles over all orientations. Since the interaction between two dipoles $V \propto 1/R^3$, and $\langle V \rangle = 0$, the statistical average over a Boltzmann distribution of the potential, $\langle V e^{-V/k_B T} \rangle$ will have a leading term proportional to $1/R^6$. A similar result had also been obtained by van der Waals (1909) by a slightly different averaging procedure. Debye (1912) used this dipole model of molecules in his theory of the dielectric constant of liquids, although his impression at that time that all molecules are polar seems to be based on his experience with those systems which are now known to have polar molecules. This was not borne out* by subsequent discoveries which revealed an absence of permanent dipole moments in the simplest molecules.

During the period extending from 1912 to 1920 an important development was the confirmation of the connection between the van der Waals

---

* The classification of substances into polar and non-polar types was introduced by G. N. Lewis (1916).

constant "a" and the index of refraction of the gas which is determined by the polarizability of the constituent molecule. Debye (1920) proposed a theory that the polarizability of molecules is the cause of molecular forces. In his treatment he assumed that each molecule has a permanent quadrupole moment, and the electric field of this quadrupole polarizes a neighbouring molecule, and the force between them is due to the quadrupole and the induced dipole. This force varies as $(1/R^9)$, and the constant of proportionality would depend linearly on the polarizability which determines the induced dipole moment. Debye then fitted the constant of proportionality with the experimental van der Waals constant "a" on the basis of a theoretically established relationship between them. Keesom (1921) indicated that Debye's assumption that each molecule has a quadrupole moment entails orientation effects in the interaction energy of two molecules. He also calculated the second virial coefficient for both dipole and quadrupole gases with each molecule having a repulsive core of diameter equal to its size. These theories, although developed at a time when knowledge of the electronic structure of molecules was inadequate and before the development of quantum mechanics, can still be considered adequate to explain the interactions between polar molecules in a phenomenological way.

The explanation of the force between a pair of non-polar molecules could come only after the advent of quantum theory. An early analysis of this problem was by Wang (1927) who used perturbation theory to solve the Schrödinger equation for two hydrogen atoms at large separation including the interactions between the electrons and protons of the two atoms, and found that the interaction energy is given by

$$V(R) = -8.7 \frac{e^2 a^5}{R^6}. \tag{1.3}$$

Eisenschitz and London (1930), and London (1930a,b) analysed the same problem and came to the same kind of result, with a factor of the order of 6.5 instead of 8.7. The subsequent developments have been in the form of more detailed analysis of the interactions taking higher moments into account (Margenau, 1931), and the effect of retardation when the distance of separation between the molecules exceeds the characteristic wavelength of radiation emitted due to dipolar transitions (Casimir and Polder, 1948). We shall consider the London and Casimir-Polder theories in some detail in Chapter 2. We shall briefly discuss here London's analysis to provide the framework for working out the force between two large dielectric materials by summing the London forces between the molecules in one and the molecules in the other.

1.2 London's theory

In its simplest form London's theory is best illustrated taking the example of two hydrogen atoms. The Hamiltonian is,

$$H = H_0 + H_I,$$

with

$$H_0 = -\frac{\hbar^2}{2m}(\nabla_1^2 + \nabla_2^2) - \frac{e^2}{r_1} - \frac{e^2}{r_2} \qquad (1.4)$$

$$H_I = \frac{e^2}{R_{12}} - \frac{e^2}{|\mathbf{r}_1 - \mathbf{R}_{12}|} - \frac{e^2}{|\mathbf{r}_2 + \mathbf{R}_{12}|} + \frac{e^2}{|\mathbf{r}_1 - \mathbf{r}_2 - \mathbf{R}_{12}|}. \qquad (1.5)$$

Here the two protons are at $\mathbf{R}_1$ and $\mathbf{R}_2$, and the electron coordinates $\mathbf{r}_1$ and $\mathbf{r}_2$ are measured from the protons (Fig. 1.2).

The effect of H_I on any of the unperturbed states of the two atoms can be studied by time-independent perturbation theory. If both the atoms are in the ground state the change in energy due to the perturbation is,

$$\Delta E = \langle 0| H_I |0\rangle + \sum_n{}' \frac{|\langle 0| H_I |n\rangle|^2}{E_0 - E_n}. \qquad (1.6)$$

If $R_{12} = |\mathbf{R}_2 - \mathbf{R}_1|$ is large compared with the Bohr radius a_0, the major contribution to the integrals in the matrix elements $\langle 0| H_I |0\rangle$ and

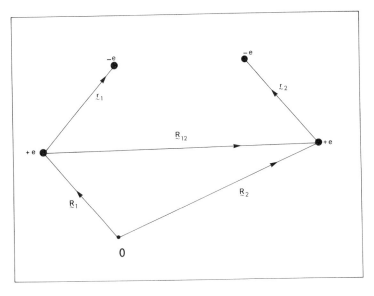

FIG. 1.2. The protons and electrons in the system of two hydrogen atoms.

$\langle 0 | H_I | n \rangle$ will come from the neighbourhood $r_j \sim a_0$, $j = 1, 2$. Hence H_I can be expressed as a multipole expansion in powers of (r_j/R_{12}) to give us in leading order,

$$H_I \cong \frac{e^2}{R_{12}^3} (x_1 x_2 + y_1 y_2 - 2 z_1 z_2), \tag{1.7}$$

where the z-axis is chosen along $\mathbf{R}_{12}$. The ground state wave function is

$$\psi_0(\mathbf{r}_1, \mathbf{r}_2) = \frac{1}{\pi a_0^3} e^{-(r_1 + r_2)/a_0}. \tag{1.8}$$

The first order correction $\langle 0 | H_I | 0 \rangle$ vanishes when H_I of eqn. (1.7) is used, and the second order term gives us,

$$\Delta E = \frac{e^4}{R_{12}^6} \sum_{m,n} \frac{|x_{0,m} x_{0,n} + y_{0,m} y_{0,n} - 2 z_{0,m} z_{0,n}|^2}{2E_0 - E_m - E_n}. \tag{1.9}$$

Here E_0 and E_n are the ground state and n-th excited state energies of the hydrogen atom, and the matrix elements $x_{0,n}$, $y_{0,n}$ and $z_{0,n}$ are taken between these states. Perturbation theoretic and variational estimates of the sum in eqn. (1.9) are well known (Schiff, 1968) and ΔE can be shown to be in the range,

$$-\frac{8e^2 a^5}{R^6} \le \Delta E \le -\frac{6e^2 a^5}{R^6}. \tag{1.10}$$

ΔE can be identified with the interaction energy $V(R)$. A careful estimate of the sum in eqn. (1.9) (Pauling and Beach, 1935) gives for $V(R)$,

$$V(R) = -6 \cdot 5 \frac{e^2 a^5}{R^6}. \tag{1.11}$$

London (1930a,b) recognized that the right-hand-side of eqn. (1.9) can be expressed in terms of the dispersion f-values or oscillator strengths. For a dipole transition between two states $|l\rangle$ and $|m\rangle$, the oscillator strength f_{lm} is defined as

$$f_{lm} = \frac{2m}{\hbar^2} (E_m - E_l) |z_{lm}|^2. \tag{1.12}$$

In eqn. (1.9) in dealing with products of the type $x_{0,m} x_{0,n} y_{0,m} y_{0,n}$ it is always possible to choose m and n such that only one of the elements $x_{0,m}$, $y_{0,m}$ and $z_{0,m}$ does not vanish. We thus get

$$V(R) = \frac{e^4}{R^6} \sum_{m,n} \frac{x_{0,m}^2 x_{0,n}^2 + y_{0,m}^2 y_{0,n}^2 + 4 z_{0,m}^2 z_{0,n}^2}{(E_0 - E_m) + (E_0 - E_n)}$$

$$= \frac{6e^4}{R^6} \sum_{m,n} \frac{z_{0,m}^2 z_{0,n}^2}{(E_0 - E_m) + (E_0 - E_n)}$$

$$= \frac{3e^4\hbar^4}{2m^2R^6} \sum_{m,n} \frac{f_{0m}f_{0n}}{(E_m - E_0)(E_n - E_0)} \frac{1}{(E_0 - E_m) + (E_0 - E_n)}. \qquad (1.13)$$

In this equation the second step is obtained from the first by symmetry considerations. This formula, although derived for the example of the hydrogen atom, is of general validity. For an atom or molecule with n electrons f_{lm} satisfies the sum rule (Landau and Lifshitz, 1958)

$$\sum_m f_{lm} = n. \qquad (1.14)$$

f_{lm} can be obtained from experimental data on intensity of spectral lines (Garstang, 1966). If we take two molecules with n_1 and n_2 electrons which have strong transitions at two frequencies ω_1 and ω_2 respectively, eqn. (1.13) reduces to the form

$$V(R) \cong -\frac{3e^4\hbar}{2m^2R^6} \frac{n_1 n_2}{\omega_1\omega_2(\omega_1 + \omega_2)}. \qquad (1.15)$$

Here the strong f-values have been equated to n_1 for one molecule and n_2 for the other, following eqn. (1.14). The form of $V(R)$ in eqn. (1.15), originally given by London, is particularly useful in estimating the dispersion force between molecules from experimental spectroscopic data.

It is possible to write eqn. (1.13) in a form that involves the polarizabilities of each molecule separately. This involves the use of the identity,

$$\frac{1}{ab(a+b)} = \frac{2}{\pi} \int_0^\infty \frac{d\xi}{(a^2 + \xi^2)(b^2 + \xi^2)}, \qquad a, b > 0 \qquad (1.16)$$

and the definition of the dynamic polarizability (Blokhintsev, 1964)

$$\alpha(\omega) = 2e^2 \sum_n \frac{(E_n - E_0)|z_{0n}|^2}{(E_n - E_0)^2 - (\hbar\omega)^2}$$

$$= \frac{e^2}{m} \sum_n \frac{f_{0n}}{\omega_{n,0}^2 - \omega^2}. \qquad (1.17)$$

Then, eqn. (1.13) becomes,

$$V(R) = -\frac{3\hbar}{\pi R^6} \int_0^\infty d\xi\, \alpha_1(i\xi)\alpha_2(i\xi) \qquad (1.18)$$

where $\alpha_j(i\xi)$ stands for $\alpha_j(\omega)$ in eqn. (1.17) with ω replaced by $(i\xi)$. Equation (1.18) enables us to evaluate the constant occurring in the $(1/R^6)$ interaction from knowledge of the dynamic polarizability. If one of the oscillator strengths in each molecule predominates over the others, from eqn. (1.17) we can write

$$\alpha_j(\omega) \cong \frac{e^2 n_j}{m(\omega_j^2 - \omega^2)}, \qquad j = 1, 2 \tag{1.19}$$

so that eqn. (1.18) becomes

$$
\begin{aligned}
V(R) &\cong -\frac{3\hbar}{2R^6}\left(\frac{e^2 n_1}{m\omega_1^2}\right)\left(\frac{e^2 n_2}{m\omega_2^2}\right)\frac{\omega_1\omega_2}{\omega_1+\omega_2} \\
&= -\frac{3\hbar}{2R^6}\alpha_1(0)\alpha_2(0)\frac{\omega_1\omega_2}{\omega_1+\omega_2} \tag{1.20} \\
&= -\frac{3\hbar}{2R^6}\frac{e}{m^{1/2}}\frac{\alpha_1(0)\alpha_2(0)}{[\alpha_1(0)/n_1]^{1/2}+[\alpha_2(0)/n_2]^{1/2}}. \tag{1.21}
\end{aligned}
$$

From eqn. (1.20) or (1.21), the interaction can be estimated in terms of the static value of the polarizability in those situations where each molecule is characterized by a strong adsorption frequency. Equation (1.21) was obtained by Slater and Kirkwood (1931).

The approximation implied in eqn. (1.19) is obviously equivalent to treating each molecule as a harmonic oscillator with a frequency ω_j and static polarizability $(e^2 n_j/m\omega_j^2)$. This oscillator model of a molecule is often very convenient for qualitative and semi-qualitative analysis of forces between molecules. A simple use of this model was, in fact, given by London (1930b) in his derivation of the $(1/R^6)$ law of dispersion forces. This model has appeared off and on in the analysis of related problems since then.

Since our object in this book is to analyse the role of dispersion forces between macroscopic bodies, we shall stop our analysis of the force between molecules at this point. The other features of intermolecular forces arising from anisotropy in the polarizability tensor of a molecule, the interaction due to induced multipole moments of higher order etc. have been adequately surveyed by Margenau and Kestner (1971). The effect of retardation will be the subject of a detailed analysis in Chapter 2. In the following sections we shall consider the application of the simple London theory of $(1/R^6)$ interaction in deriving the dispersion force between macroscopic bodies. The interaction due to induced multipole moments of higher order, which correspond to a variation of $V(R)$ as $1/R^{2n}$ with $n > 3$, and the effect of anisotropy of the molecular polarizability will not be considered in this analysis, since the major contribution comes from the dipolar forces.

1.3 The method of pairwise summation

The existence of some long range force between bodies of colloidal dimensions can be inferred from the very existence of the phenomenon of flocculation. Credit for the recognition that such a force would emerge from summing the London–van der Waals forces between pairs of atoms of the interacting particles appears to be due to Kallmann and Willstätter (1932), and to Bradley (1932). The former paper is interesting besides, since it contains the first explicit suggestion that a complete picture of the force field between colloidal bodies could be obtainable on the basis of the repulsive double-layer (electrostatic) interactions combined with the attractive dispersion forces. The suggestion was taken up subsequently by Verwey and Overbeek (1948) and by Deryaguin and Landau (1941) who developed independently the theory of colloid stability which has remained unchallenged for 30 years. (Landau reports that Ellis (1912) reported that Donnan had suggested this much earlier.) Bradley's paper is also of historical interest; for, stimulated by earlier experiments of Tomlinson (1928, 1930), he first calculated the force between two spheres, and then went on to report a remarkably simple experiment which provided a direct measurement of the cohesive energy of macroscopic spheres. While crude, this experiment anticipated by some 40 years the more refined and definitive work done recently (Israelachvili and Tabor, 1972a, 1972b, 1973).

Subsequently, de Boer (1936) and Hamaker (1932) investigated theoretically the dispersion forces acting between colloidal objects.* Hamaker considered spherical bodies, assumed pairwise additivity of interatomic dispersion energies, and demonstrated the essential result—that although the range of atomic forces was of order of atomic dimensions, the sum of the dispersion energies led to an interaction range for colloidal bodies of the order of their dimensions. The work of Hamaker and the many early papers of de Boer and his collaborators have played a central role in subsequent estimates of interaction energies which are fundamental in colloid science. Like most simple theories the de Boer–Hamaker approach to interactions has the virtue not only of ease in comprehension, but works over a wider range than would at first be thought possible. We now consider this approach in more detail.

---

* Strictly speaking, although he did not consider the interaction problem directly, a student of the history of science would probably assign the method of pairwise summation of potentials to Gauss who first developed the technique in connection with his investigations of Laplace's theory of Capillary Action; see Maxwell (1875).

1.4 Results for different geometries

Consider then the energy of interaction between macroscopic bodies which have well-defined geometric shapes and uniform densities* so that integration over their volumes can be performed. Assume further that the polarizabilities of the constituent molecules of the interacting bodies are isotropic, and that the bodies are separated by a vacuum. The interaction energy of two molecules due to dispersion forces, as we have seen in the previous sections, can be written as

$$E(r) = -\frac{\Lambda_{12}}{r^6}, \tag{1.22}$$

where the constant is obtained from eqn. (1.21). Granted additivity, and that the bodies contain molecules at densities ρ_1, ρ_2, the whole energy of interaction will be given by

$$E^A = -\int_{V_1} d\tau_1 \int_{V_2} d\tau_2 \frac{\Lambda_{12}\rho_1\rho_2}{r_{12}^6}, \tag{1.23}$$

where r_{12} is the distance between $d\tau_1$ and $d\tau_2$; $d\tau_1$, $d\tau_2$ denote volume elements of bodies 1 and 2 with volumes V_1, V_2. Notice that the whole energy splits into a product of two terms, $\rho_1\rho_2\Lambda_{12}$, and another which depends only on geometry. This separation is a consequence of the assumption of pairwise summation, which is not valid in general.

When stated in the above terms, the problem appears simple, but to obtain results in closed analytic forms is extremely difficult except for the most elementary geometries, which we now consider.

Two slabs

The situation pictured in Fig. 1.3 serves as a model for, say, two inter-acting clay plates; or for the two hydrocarbon layers (separated by water

---

* The assumption of uniform density near the surface of a liquid is a question which has beset colloid and interface science since Poisson (1831) first raised the issue in his theory of surface energies which maintained the opposite. The problem is still unresolved. The assumption of uniform density up to the interface is probably fair. For most applications this is almost certainly so. On the other hand if we consider the extreme case of the equilibrium between a gas and its liquid near the critical point, it is perfectly clear that the width of the transition region over which densities can vary can become very broad. the papers of Widom (1965) on this topic are especially worth reading. The methods of Poisson could probably be invoked with advantage in studying this problem, which has recently been readdressed by Triezenberg and Zwanzig (1972). It is only in the past two or three years that experimental techniques have become available for the investigation of interface profiles in rare gas liquids near their critical points and the new art of ellipsometry promises to shed some light on the density profile of adsorbed layers. For the moment we accept the assumption of uniform density with infinitely sharp discontinuity at the surface.

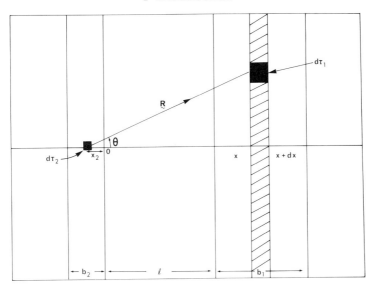

FIG. 1.3. Volume elements in the two slab system for use of formula (1.23).

and bounded by air), of a model soap film; or for two biological cells in close proximity, provided we ignore the effect of the intervening aqueous medium, and consider situations in which the separation distance is small, so that the curvature of the surfaces can be neglected. By symmetry the force between the two plates is in the direction of the x-axis. Differentiating (1.22) we have for the force between two molecules,

$$F(R) = -\nabla E(R) = -\frac{6\Lambda_{12}}{R^7} \tag{1.24}$$

and the force on an elementary volume $d\tau_2$ due to the molecules in $d\tau_1$ will be

$$F_x = -\rho_1\rho_2 \frac{6\Lambda_{12}}{R^7} \cos\theta. \tag{1.25}$$

The volume of the elemental slab between x and $x + dx$ in polar coordinates is $2\pi r\, dr\, dx$, with $\cos\theta = (x - x_2)/R$, $R^2 = r^2 + (x - x_2)^2$, whence for the force on $d\tau_2$ we have

$$\begin{aligned}
F_x &= -12\pi\rho_1\rho_2\Lambda_{12}\int_0^\infty r\, dr \int_l^{l+b_1} \frac{(x - x_2)\, dx}{[r^2 + (x - x_2)^2]^4} \\
&= -\frac{\pi\rho_1\rho_2}{2}\Lambda_{12}\left(\frac{1}{(l - x_2)^4} - \frac{1}{(l + b_1 - x_2)^4}\right).
\end{aligned} \tag{1.26}$$

The total force on unit area of slab 2 will then be

$$F^A = -\frac{\pi\rho_1\rho_2}{2}\Lambda_{12}\int_{-b_2}^{0}dx_2\left(\frac{1}{(l-x_2)^4}-\frac{1}{(l+b_1-x_2)^4}\right)$$

$$= -\frac{\pi\rho_1\rho_2}{6}\Lambda_{12}\left(\frac{1}{l^3}-\frac{1}{(l+b_2)^3}-\frac{1}{(l+b_1)^3}+\frac{1}{(l+b_1+b_2)^3}\right). \quad (1.27)$$

The attractive potential follows by integration with respect to l from l to ∞:

$$E^A = -\frac{\pi\Lambda_{12}\rho_1\rho_2}{12}\left(\frac{1}{l^2}+\frac{1}{(l+b_1+b_2)^2}-\frac{1}{(l+b_1)^2}-\frac{1}{(l+b_2)^2}\right). \quad (1.28)$$

By convention the quantity

$$A_{12} = \pi^2\rho_1\rho_2\Lambda_{12} \quad (1.29)$$

is called the Hamaker constant. We shall return to further discussion of this result in our analysis of soap film experiments. Note the special cases

(i) $b_1, b_2 \to \infty$

$$E^A = -\frac{A_{12}}{12\pi l^2}. \quad (1.30)$$

The force per unit area between two semi-infinite plates is $F^A = -(A_{12}/12\pi l^3)$. Typically, from (1.29) and (1.19), using the Lorentz formula linking polarizability with refractive index, viz. $\varepsilon - 1 = 4\pi\rho\alpha$, we have

$$A_{11} = \frac{3\pi^2}{4}\hbar\omega_0\rho_1^2\alpha_1^2(0) \approx \frac{3}{64}(\varepsilon-1)^2\hbar\omega_0. \quad (1.31)$$

For water, or glass, $\varepsilon \approx n^2 \approx 1.7$, where n^2 is the square of the refractive index. Also, a typical characteristic (ultraviolet) absorption frequency is $\omega_0 \sim 10^{16}$ rad/sec, and

$$F^A \sim -\frac{3\times10^{-12}}{4l^3}\ \text{dynes/cm}^2, \quad (1.32)$$

where l is measured in cm. When an intermediate liquid separates the planes the characteristic value $A \sim 10^{-12}$ for the Hamaker constant appropriate to a vacuum can decrease by as much as two orders of magnitude.

(ii) $b_1 = b_2 = b$

$$E^A \approx -\frac{A}{12\pi l^2}\left\{1-\frac{7}{4}\left(\frac{l^2}{b^2}\right)+0\left(\frac{l}{b}\right)^4\right\}, \qquad b > l$$

$$\approx -\frac{A}{2\pi l^2}\left(\frac{b}{l}\right)^2\left(1-\frac{4b}{l}+\cdots\right), \qquad b \ll l. \quad (1.33)$$

A point to be noted is that the energy becomes infinite at zero separation, so that an infinite amount of work has to be done in creating a surface by pulling the plates apart. But this work, which is just twice the surface energy is certainly finite. Thus, this form of the interaction energy leads to infinite surface energy. In this connection it is worth noting that were we to extrapolate (1.33) down to a distance l of the order of an atomic diameter, the energy E^A so computed would be of the same order of magnitude as measured surface energies* (Israelachvili, 1973).

Multilayers

The extension of the above analysis to a multilayer of plates is immediate (Fig. 1.4). Again we ignore the presence of an intervening liquid. Such a system occurs frequently in biological situations, for example in lecithin figures, myelin sheaths, within cells generally, in the so-called neat phase of soap solutions or lamellar liquid crystals, and in considerations on the flocculation of clays. The results are: For an infinite multilayer of plates of thickness b, separation l, the energy of formation per plate again measured with respect to a zero of energy at infinite separation, is

$$E^A = -\frac{M_6}{12\pi l^2},$$

$$M_6 = A \sum_{j=0}^{\infty} \left\{ \frac{1}{\left\{ \left(1+\frac{b}{l}\right)j+1 \right\}^2} - \frac{2}{\left\{ (j+1)\left(1+\frac{b}{l}\right) \right\}^2} \right.$$
$$\left. + \frac{1}{\left\{ (j+1)\left(1+\frac{b}{l}\right)+\frac{b}{l} \right\}^2} \right\} \tag{1.34}$$

* It would be astonishing if this were not so, for to suppose otherwise would imply that the attractive forces between atoms responsible for the cohesion of a solid or liquid were different from those which give rise to surface tension. This was already realised by Young (1805) in his essay on the cohesion of fluids. Remarkably, many years before the general acceptance of the idea of atoms by the chemists, Young in his usual inimitable style had given estimates of the distance of atoms in a simple liquid and the range over which their attractive forces act. His estimate of both was of the order of angstroms which of course is correct. It was based simply on the strength of materials as compared to surface energies, and was practically forgotten for a hundred years, mainly because of Young's too scrupulous avoidance of mathematical symbols. (Young is somewhat scathing about Laplace whom he practically accuses of stealing his theory verbatim.) Nobody paid any attention to it till Rayleigh (1890) redressed the balance. With the exception of Langmuir (1916, 1917), van Urk (1937), and a few others, it seems to have been forgotten again in this century. The early work is fascinating to read, as also is the essay of J. Clerk Maxwell (1875) on Capillary Action in the 9th edition of the Encyclopaedia Britannica. The last includes a complete history of the subject of surface energies till that date.

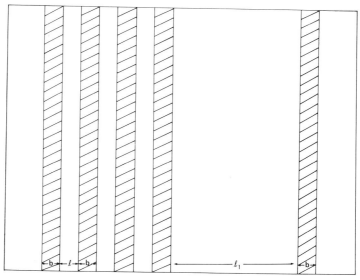

FIG. 1.4. The multilayer system.

with limiting approximations

$$M_6 \approx A\left\{\frac{1}{2}+\frac{1}{\left(1+\dfrac{b}{l}\right)}-\frac{\pi^2}{6\left(1+\dfrac{b}{l}\right)^2}+\cdots\right\}; \qquad l<b \qquad (1.35a)$$

$$\approx \frac{Ab^2}{l^2}\left\{3+\frac{2l}{l+b}\right\}; \qquad\qquad\qquad l\gg b. \qquad (1.35b)$$

If one applies the same method to find the energy of a single layer at a distance l_1 from an already formed multilayer stack, corresponding to (1.33) and (1.35b), we have

$$E^A \approx -\frac{A}{12\pi l_1^2}\left\{\frac{b^2}{l_1^2}\right\}\left\{\frac{2l_1}{l+b}+3+0\left(\frac{l+b}{l_1}\right)\right\}, \qquad (1.36)$$

which is of the order of $[2l_1/(l+b)]$ larger than the corresponding result (1.33) for two thin plates. For large distances l_1 we might expect the left hand plate to "see" the stack as a continuous medium so that this result is not altogether surprising. From a different point of view though, it is. For, the equation tells us that a stack, once formed, will tend to increase the number of its layers because the range of its attractive force increases substantially with the number of its members. The application of these results, and their success or failure in several situations will be mentioned further in Chapter 5.

Spheres

Pairwise summation for spheres also presents no particular problem (Hamaker, 1937). Consider a point P at distance $OP = x$ from the centre of a sphere of radius b_1 (Fig. 1.5). The surface ABC, S_{ABC}, cut out from a sphere of radius r around P by the sphere around O is

$$S_{ABC} = 2\pi \int_0^{\theta_0} r^2 \sin \theta \, d\theta, \tag{1.37}$$

where θ_0 is determined by

$$b_1^2 = x^2 + r^2 - 2xr \cos \theta_0. \tag{1.38}$$

Thus

$$S_{ABC} = \frac{\pi r}{x}[b_1^2 - (x - r)^2] \tag{1.39}$$

and since $d\tau_1 = S_{ABC} \, d\tau$, the potential energy of interaction of an atom at P due to a spherical particle with its centre at O is

$$E_p = -\int_{x-b_1}^{x+b_1} \frac{\rho_1 \pi \Lambda_{12}}{r^6} \frac{r}{x}[b_1^2 - (x - r)^2] \, dr, \tag{1.40}$$

whence from (1.23) the total energy of interaction between two spheres

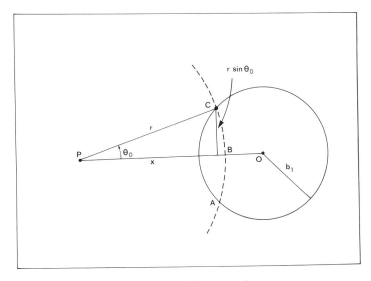

$r \sin \theta_0$

FIG. 1.5. A particle near a sphere.

with their centres a distance R apart is

$$E^A = \int_{R-b_2}^{R+b_2} E_p \rho_2 \pi \frac{x}{R} [b_2^2 - (R-x)^2] \, dx. \tag{1.41}$$

The result for two spheres of radii b_1, b_2 is

$$E^A = -\frac{A_{12}}{6} \left\{ \frac{2b_1 b_2}{R^2 - (b_1 + b_2)^2} + \frac{2b_1 b_2}{R^2 - (b_1 - b_2)^2} + \ln \left(\frac{R^2 - (b_1 + b_2)^2}{R^2 - (b_1 - b_2)^2} \right) \right\}. \tag{1.42}$$

For spheres of equal size

$$E^A = -\frac{A_{12}}{6} \left\{ \frac{2b^2}{R^2 - 4b^2} + \frac{2b^2}{R^2} + \ln \left(1 - \frac{4b^2}{R^2} \right) \right\}. \tag{1.43}$$

The limiting form for surface-to-surface separation $l = R - 2b \ll b$ is

$$E^A \approx -\frac{A}{12} \frac{b}{l} \left\{ 1 + \left(\frac{3}{8} + \ln \left(\frac{l}{b} \right) \right) \frac{2l}{b} + \cdots \right\} \tag{1.44}$$

while for large distance the energy decays as $1/R^6$.

The result, eqn. (1.44), is peculiar. As remarked by Verwey and Overbeek (1948);

> "The energy of attraction decays very slowly, i.e., reciprocally with distance, even slower than it does in the case of flat plates. At larger separations the decay is faster ($\frac{16}{9} A_{12} b^6 / (R - 2b)^6 \sim 1/R^6$) but the decay remains slower than the planar case ($\propto 1/l^2$) until $R/b \gg 2.4$. This slow decay is understandable by the fact that for $R/b \sim 2$, only small parts of the spheres are very close together. A small change of the distance between the spheres has a substantial effect only on the elements surrounding the points of closest approach, whereas for the bulk of the material involved in the interaction the relative change of the distances, and the consequent change in attractive energy are small."

This model cannot be applied realistically to two spheres at close proximity, because in real situations there will be some distortion from spherical shape at close separation due to the effect of the interaction. This distortion in shape may become of increasing importance for liquid drops or other bodies of low cohesive energy.

Sphere and a plane

This case follows from (1.42) by letting one of the radii go to infinity. The result is (R_1 is here the distance from the sphere centre to the plane)

$$E^A = -\frac{A_{12}}{6} \left(\frac{1}{x} + \frac{1}{2+x} + \ln \frac{x}{2+x} \right), \tag{1.45}$$

where $x = [(R_1 - b_1)/b_1]$. For $x \ll 1$, $E^A \rightarrow -(A/6)b_1/l$, which is twice the value for two spheres of equal radius. For a single atom at a large distance l from a plane

$$E^A \sim -A_{12}\left[\frac{2}{9}\left(\frac{b_1}{R_1}\right)^3 + \frac{4}{15}\left(\frac{b_1}{R_1}\right)^5 + \cdots\right]. \tag{1.46}$$

The formulae (1.42) and (1.45) illustrate another point, that the interaction energy is determined mainly by the size of the smaller of the interacting particles. [The correction terms to eqn. (1.45) involve powers of b_1/b_2 and the radius b_2 disappears in leading order.] Another interesting result due to Hamaker (1937) emerges from formula (1.42) if we observe that the interaction energy is a function of the ratios b_1/b_2, R/b_1 or R/b_2. This means that if we make each particle K times as large, and put them K times apart, the energy retains the same value. This can be understood in a simple way which leads to a fundamental generalization. For, suppose the interaction energy between two atoms is $\varepsilon = -(\Lambda/r^q)$. The interaction energy between volume elements $d\tau_1$, $d\tau_2$ of the separate particles is

$$dE^A = -\Lambda\rho^2 \frac{d\tau_1 \, d\tau_2}{r^q}. \tag{1.47}$$

If all geometric dimensions are increased by a factor K, the contribution of the corresponding volume elements $d\tau_1' \, d\tau_2'$ in the new configuration will be

$$dE'^A = -\Lambda\rho^2 \frac{d\tau_1' \, d\tau_2'}{(r')^q} = -\Lambda\rho^2 K^3 \frac{d\tau_1 K^3 \, d\tau_2}{K^q r^q} = \frac{dE^A}{K^{q-6}}. \tag{1.48}$$

In other words, if we change the size of the particles by K and put them K times apart, the energy of interaction changes as $1/K^{q-6}$. The energy remains constant when $q = 6$, decreases when $q > 6$, and increases when $q < 6$. Thus the force of gravity which is negligible when acting between atoms or colloidal particles dominates in celestial mechanics. This was known to Newton.

Cylinders and other shapes

In addition to spheres and planes, the integral of eqn. (1.23) has been evaluated for other figures, in particular for circular discs, both in co-planar and sandwich-like configurations, and for spherical shells. Vold

(1954, 1957) has considered approximately certain rectangular parallelepipeds and ellipsoids,* and de Rocco and Hoover (1960) have considered collinear thin rods, parallel rods, co-planar rectangles, rectangles in sandwich configurations, aligned and skew parallelepipeds. The last mentions references to earlier calculations as do the papers of Salem (1962) and Sparnaay (1959). The review of Israelachvili (1974) contains a number of references to more recent work. Most of these papers are naturally concerned with more than the problem of integrating $1/r^6$. Sparnaay (1959) following work of Bouwkamp (1947), considered the difficult problem of parallel cylinders and cylinders inclined at a right angle. In the former case one cylinder is infinite, and in the latter both are infinite. Much of this earlier work, while providing a useful qualitative guide to the effects of geometry on van der Waals forces has been superseded. The statement is true for even the most recent papers (Brenner and McQuarrie, 1973) which attempt to examine the force balance between cylinders and cylindrical arrays. This is particularly so in biological applications, where nature conspires to limit the applicability of elementary models. Three-body and higher order interactions, nonadditivity of forces, anisotropy, the interplay between double-layer and electromagnetic fluctuation forces, and a number of other effects form only a fraction of the armoury of the conspirator. Interactions between cylindrical bodies are expected to play a key role in the study of soaps and detergents, lubricants and associated surface phenomena, in the theory of liquid crystals and—in biology, to mention a few applications—in the stabilization of tobacco mosaic virus, of interactions in muscle proteins, helical assemblies in aqueous solution and in considerations involving microtubules and cilia.

The main points which can emerge from the method of pairwise summation were already exhibited by Vold (1954, 1957) and are: (1) that long thin structures will tend to align themselves; (2) that rectangular rods

---

* M. J. Vold, under the direction of Overbeek, appears to have been the first to recognize the role of shape in determining the nature of flocculation phenomena, and the two very simple clear papers (Vold, 1954, 1957) underline this point of view. Herbert Jehle, while focusing attention on material properties as being the origin of specificity in biological macromolecule interactions had also recognized the possible importance of shape in determining long range specificity. His views, promulgated with vigour in a series of papers through the 1950s were noted and largely ignored by physicists and biologists until the matter was taken up seriously by Parsegian (1973) and others a few years ago. That van der Waals forces might be regarded as a cause of specificity and self-assembly in biology appears to have been well-recognized by the British surface and colloid chemists, especially Bangham, Rideal, and Pethica, and also by Fröhlich and Longuet–Higgins among physicists. Considerable credit is also due to Parsegian (1973) who set out to properly define the problem.

will align so that their largest faces are opposite, and a key point (3) that the total attractive dispersion energy between two colloidal particles is greater than or of the same order of magnitude as thermal energies (and hence significant) when the mean diameter ($V^{1/3}$ where V is the particle volume) is of the order of particle separation.* This conclusion holds regardless of whether the particles are spherical, rod- or plate-shaped. The relative order of attractive energies is: plates>rectangular-rods> cylinders>spheres. This result is not unexpected if we observe that for surface energies at fixed volume the inequalities run in the opposite direction.

Cylinders

There are two forms for the interaction between infinitely long parallel cylinders useful for computation. Both are equally obnoxious.

If R is the centre-to-centre distance and b is the radius of each cylinder it is straightforward to reduce eqn. (1.23) to a double integral in the manner indicated by Elliott (1968) to find the energy of interaction per unit length as

$$E^A = -\frac{3}{2\pi} A \int_{R-b}^{R+b} dx \int_{x-b}^{x+b} dy \frac{x}{y^4} \cos^{-1}\left(\frac{R^2+x^2-b^2}{2Rx}\right) \cos^{-1}\left(\frac{R^2+y^2-b^2}{2xy}\right).$$

(1.49)

The y integral can be carried out in terms of elliptic integrals. Final expansions of the remaining integrals are straightforward but tedious. A more elegant expression is that of Bouwkamp (see Sparnaay, 1959) who obtains

$$E^A = -\tfrac{3}{8}\pi b^4 A U_5, \qquad U_5 \equiv \frac{1}{9}\left(\frac{d^2}{dR^2} + \frac{1}{R}\frac{d}{dR}\right)U_3.$$

(1.50)

The function U_3 is the square of an elliptic integral

$$U_3 = \frac{16R}{\pi^2 b^4} k^4 \left(\int_0^{\pi/2} \frac{\cos^2\phi\, d\phi}{\sqrt{1-k^2\sin^2\phi}}\right)^2$$

(1.51)

with the modulus defined by

$$k^2 = \left(1 - \sqrt{1-\frac{4b^2}{R^2}}\right)^{1/2}.$$

(1.52)

---

* This observation when combined with the short-range nature of double-layer forces could be used to make a theory of flocculation which parallels the liquid gas transition as given by the van der Waals equation of state.

The function U_5 can be written as a generalized hypergeometric function

$$U_5 = \frac{1}{R^5} F_4\left(\frac{5}{2}, \frac{5}{2}; 2, 2; \frac{b^2}{R^2}, \frac{b^2}{R^2}\right)$$

$$= \frac{1}{R^5} \sum_{m,n=0}^{\infty} \frac{\Gamma(\frac{5}{2}+m+n)}{\Gamma(\frac{5}{2})m!\, n!\, (m+1)!\, (n+1)!} \left(\frac{b^2}{R^2}\right)^{m+n}. \qquad (1.53)$$

For $R \gg b$, we have

$$U_5 = \frac{1}{R^5}\left\{1 + \frac{25}{4}\frac{b^2}{R^2} + 31 \cdot 90\frac{b^4}{R^4} + 150 \cdot 7\frac{b^6}{R^6} + \cdots\right\}. \qquad (1.54)$$

We have quoted this result at some length to illustrate the point that summation of $1/r^6$ over two volumes is much harder than it looks. Mathematically, as will be indicated in the next chapter, it is equivalent to finding the Green function for the Laplacian operator with Neumann boundary conditions, for which problem there are no systematic procedures in general. Even for relatively simple geometries there is little virtue in expressing the interaction energy "exactly," i.e., in terms of known transcendental functions. Even if this can be done, apart from the crudity of the physical model, the results will always turn out to involve functions whose convergence as a function of the relevant parameters is rather slow. Except for limiting cases, where perturbation techniques can be employed to advantage, and where the analytic results are instructive, it is much more useful to resort to numerical computation.

At small separations $l = (R - 2b) \ll b$, expansion of the elliptic integral of eqn. (1.51) followed by the required differentiations gives for the interaction energy per unit length

$$E^A = -\frac{A}{24b}\left(\frac{b}{l}\right)^{3/2}\left\{1 - \frac{l}{b} + \frac{1}{\sqrt{2\pi}} \ln\left(\frac{l}{b}\right) + \cdots\right\}. \qquad (1.55)$$

The transition from the strong ($1/l^{3/2}$) to the weak $\propto 1/R^5$ form occurs at $R/b \sim \frac{1}{4}$ [compare with spheres]. At distances very large compared with the length of the cylinder, L, the approximation of infinite length breaks down, and the interaction goes over to the $1/R^6$ form. The result is elementary:

$$E^A \approx -A\frac{b^4 L^2}{R^6}\left(1 - \frac{1}{2}\frac{L^2}{R^2} + \cdots\right); \qquad L < R. \qquad (1.56)$$

Crossed cylinders

The energy of cylinders inclined at an angle of 90° has been obtained by Bouwkamp and studied in detail by Sparnaay (1959). The interest here is in attempting to understand the relation between anisometry and flocculation properties of suspensions. Bouwkamp's result is

$$E^A = -\frac{A}{3}\left(\frac{(1-\frac{1}{2}k^2)}{(1-k^2)} E(k) - K(k)\right), \tag{1.57}$$

where $E(k)$ and $K(k)$ are complete elliptic integrals in standard notation, and where the modulus $k = (2b/R)$. The expansion in terms of (b/R) of this equation is

$$E^A = -\frac{\pi}{2} A\left(\frac{b}{R}\right)^4\left\{1 + \frac{5b^2}{R^2} + 21 \cdot 875 \frac{b^4}{R^4} + \cdots\right\}. \tag{1.58}$$

This energy is finite for infinitely long cylinders, as opposed to the case of parallel cylinders (cf. eqn. (1.54)). Comparison of the two results shows that in the absence of other forces parallel alignment of cylinders of length L is strongly favoured for $R \ll L$. (For electrostatic effects and the only successful treatment of the phase transition in suspensions of cylindrical colloidal particles see Onsager (1942, 1944).) At very close separation, $l = (R - 2b) \ll b$, expansion of the elliptic integrals gives

$$E^A = -\frac{A}{6}\frac{b}{l}\left\{1 - \frac{3}{2}\frac{l}{b} + 0\left(\left(\frac{l}{b}\right)^2 \ln \frac{l}{b}\right)\right\}, \tag{1.59}$$

so that two crossed cylinders behave like a sphere and a plane. At any arbitrary angle the result is much more difficult to obtain. This angle-dependent potential, necessary to study the statistical mechanics of long thin needle-like molecules, requires that both cylinders be finite. A useful result is given by Mitchell and Ninham (1973).

1.5 Vold's calculations

We return at a later stage to further developments which take into account other effects on the interaction of anisometric particles. Before leaving the topic, some calculations of Vold (1957) deserve mention. Vold was concerned here with the phenomenon of mechanically reversible coagulation which occurs in suspensions of metal soaps in hydrocarbons (lubricating greases) and in which seemed not to square with the accepted picture of flocculation then current. The particular applications need not concern us, but the conclusions of her calculations are of much general interest. Illustrative of the effects of geometry is Fig. 1.6. Similar curves hold for oblate spheriods, and for anisometric parallelepipeds. In application to suspensions the interactions at constant separation of centres of mass are of importance in considering the effect of concentration on flocculation properties. For a given separation of centres of mass,

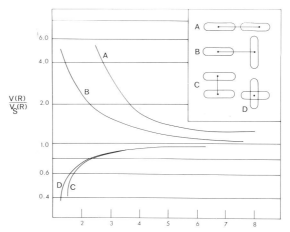

FIG. 1.6. Interaction energy $V(R)$ of prolate spheroids with semi-axis ratio $a/b = 10$ relative to spheres at the same separation of centres of mass in various orientations. The inset shows particle orientations. [Taken from Vold (1957).]

the interaction energy is *largest* for end-to-end orientation of needles for lath-shaped bodies and edge-to-edge orientation of plates. Hence the majority of particles will have this orientation when their Brownian motion brings them into close proximity with each other. The same conclusion holds for isometric particles (like micelles). For if two isometric particles cohere, the resulting aggregate is necessarily anisometric, and hence will tend to orient itself "end-on" to a third particle with increasing density of particles. Such chain-like aggregates of spherical particles are often observed.* Although primitive, such considerations throw light on the phase transitions observed in soap solutions, and provide a first hint towards understanding of biological self-assembly.

1.6   The effect of an intervening substance

Our considerations so far have been limited to situations where the particles are separated by vacuum. If the particles are immersed in a fluid, the interaction must clearly be modified, and at first sight one might think that if the attraction between atoms of the particle and those of the fluid is greater than that between the particles themselves, the net force will be

---

* The observation will not be surprising to the dedicated beachcomber. While the patterns of flotsam on a calm day have not been analysed as a function of the driving frequency of shore waves the accumulations are much analogous to those observed in colloid science, and probably have the same mathematical origin. An electric sanding machine, pressed onto a hard bench, will induce the same ordering processes among small slabs of wood.

a repulsion rather than an attraction. This is not so. The argument which leads to this conclusion is due to de Boer and was first reported by Hamaker (1937). The conclusion is so fundamental that it is worth stating as the "de Boer–Hamaker theorem":*

"The London–van der Waals forces between two particles of the same material embedded in a fluid is always attractive, provided the fluid is structureless and the particles are constrained to have a fixed size and shape. If the particles have a different composition, the resultant force can be a repulsion." The argument which leads to the statement as given by Hamaker (1937) is as follows.

Consider two bodies S and T consisting of two different solid substances 1 and 2 embedded in a fluid 0. In evaluating the energy variations we have to take into account not only the bodies S_1 and T_2 but also bodies of the same size consisting of the fluid 0, i.e., the displaced volumes of the fluid. Denote these as S_0 and T_0. Let

E_{12} = energy of interaction of S_1 and T_2
E_{10} = energy of interaction between the solid S_1 with the fluid body T_0
E_{20} = energy of interaction between the fluid body S_0 and the solid T_2
E_{00} = energy of interaction between the two fluid bodies S_0 and T_0.

These energies will be functions of the distance between the bodies. Now if E_1 represents the energy of the solid S_1 in the liquid at infinity, this body when brought to the neighbourhood of the body T_2 will have energy $E_1 + E_{12} - E_{10}$. But while bringing S_1 towards T_2 we have at the same time to remove the fluid body S_0 towards infinity. To this will correspond a change in energy from $E_0 + E_{20} - E_{00}$ to E_0 where E_0 is the energy of S_0 at infinity. Since E_1 and E_0 are constants the energy changes attendant to variations in the distance between the bodies S_1 and T_2 will be

$$E = E_{12} - E_{10} - (E_{20} - E_{00}) = E_{12} + E_{00} - E_{10} - E_{20}. \qquad (1.60)$$

This expression as yet is independent of the nature of the force. Hamaker's argument however depends on the assumption that the energy of interaction of one solid body with the fluid shall be unaffected by the presence or absence of the other solid body. This limitation will be violated as soon as the interaction between the solid bodies and the fluid is accompanied by an orientation of the fluid molecules, for the degree of orientation with respect to T_2 will certainly be affected by the presence of

---

* The de Boer–Hamaker theorem is closely analogous to the phenomenon whereby two like particles floating on a liquid always attract—independently of whether the liquid wets the particles or not. If one particle is wetted by the liquid and the other not, the floating corks problem becomes much more complicated, and the interaction can be attractive at some distances, repulsive at others.

the body S_1 (e.g., in a dipolar fluid like water, water molecules will tend to cluster around S_1 with a preferred average orientation). When T_2 is close this orientation will be altered and causes a change of energy. Besides this effect, we recall that London forces are due to the mutual polarization of atoms. If the forces are strong, the induced polarization with respect to say an atom 1 and another 0 will be affected by neighbouring atoms of 1 or atoms of 2, i.e., three-body and other many-body forces would matter; but we shall ignore such effects.

On the basis of the two assumptions we can now investigate the energy between two spherical particles by way of an example. The energy in vacuo of two like spheres is

$$E^A = -Af(G), \tag{1.61}$$

where $f(G)$ is a functional of geometry and $A = \pi^2 \rho^2 \Lambda$, ρ being the number of atoms/cm^3. If the two particles are different we have by (1.23), (1.21)

$$A_{12} = \pi^2 \rho_1 \rho_2 \Lambda_{12}, \qquad \Lambda_{12} = \tfrac{3}{2} \alpha_1 \alpha_2 \frac{\hbar \omega_0^{(1)} \omega_0^{(2)}}{(\omega_0^{(1)} + \omega_0^{(2)})}. \tag{1.62}$$

The form of Λ_{12} given in this equation is consistent with the oscillator model of the constituent molecules discussed in §1.2. Here $\omega_0^{(j)}$ is the principal absorption frequency of the j-th species. Then eqn. (1.60) implies that E^A as given by eqn. (1.61) is still valid provided we assign to A the effective value

$$A_{12}^{(\text{eff})} = \pi^2 \{ \rho_1 \rho_2 \Lambda_{12} + \rho_0^2 \Lambda_{00} - \rho_0 \rho_1 \Lambda_{01} - \rho_0 \rho_2 \Lambda_{02} \}. \tag{1.63}$$

Substituting eqn. (1.62) into this expression and writing

$$\mu_i = \alpha_i \rho_i I_i, \qquad I_i = \hbar \omega_0^{(i)},$$

we have

$$A_{12}^{(\text{eff})} = \frac{3\pi^2}{2} \left\{ \frac{\mu_1 \mu_2}{I_1 + I_2} + \frac{\mu_0^2}{2I_0} - \frac{\mu_0 \mu_1}{I_0 + I_1} - \frac{\mu_0 \mu_2}{I_0 + I_2} \right\}. \tag{1.64}$$

For two bodies of the same substance 1,

$$\begin{aligned}
A_{11}^{(\text{eff})} &= \frac{3\pi^2}{2} \left\{ \frac{\mu_1^2}{2I_1} + \frac{\mu_0^2}{2I_0} - \frac{2\mu_0 \mu_1}{I_0 + I_1} \right\} \\
&= \frac{3\pi^2}{2} \{ I_1 I_0 (\mu_0 - \mu_1)^2 + (\mu_0 I_1 - \mu_1 I_0)^2 \} > 0. \tag{1.65}
\end{aligned}$$

Hence A_{11} will be positive, which means that the force is always attractive. In general, for bodies of different substances the force can be either positive or negative. The result can be extended to arbitrary bodies of mixtures of different atoms, and if the two bodies have the same composition, attraction always results. The theorem can also be extended to arbitrary layered media, and rules of combination to form effective Hamaker constants worked out. There has been much debate in the literature concerning the correctness or otherwise of keeping only two-body effects, and the conventional prescription suggested by Verwey and Overbeek (1948) has been to divide A_{11} by the dielectric constant of the intervening medium in the visible range. Since the older methods have been superseded and sometimes lead to quite incorrect answers we do not discuss these modifications further here. The older methods of calculating Hamaker constants are clearly described in the useful paper of Gregory (1969), who gives also an extensive list of references to earlier work.

1.7 Difficulties in the method of pairwise summations

We have already alluded to some of the difficulties inherent in the method of pairwise summation. These difficulties in the main have been clear to workers in the field from the beginning, but it is only in the last couple of decades that experiment and theory have advanced to a stage where the limitations of the method have to be tackled. A partial list of the limitations is as follows.

(i) The method of pairwise summation of interactions of the constituent molecules must fail to some extent for condensed media, in which the three-body and other many-body forces must contribute. No account is taken of this aspect.

(ii) To calculate Hamaker constants it is usual to assume, as has been done here, that contributions to the London–van der Waals force are centred around a single dominant frequency (usually in the ultraviolet). This is not a realistic representation of the actual optical behaviour of molecules. The molecular polarizability as a function of frequency can be determined accurately with modern spectroscopic techniques, and this must be taken into consideration in computation of Hamaker constants (such as, through use of eqn. (1.18)).

(iii) In treating the constituent molecules as point particles, as has been done customarily in this method, certain spurious divergences arise in the theory. One such, for instance, is the divergence in the force between slabs at zero separation. Such divergences can be removed by a systematic analysis of atomic size effects on dispersion energies, and will be discussed in detail in Chapter 4.

(iv) The assumption that bodies have fixed shapes and sizes, and that they suffer no distortion in shape when they are in close proximity, is

clearly unrealistic. In actual fact they must undergo distortion in shape
due to the interactions.

A proper form of the theory of dispersion interactions must tackle the
limitations, not only of the method of pairwise additivity of dispersion
forces, but also of the microscopic approach in general. It must take into
account many-body effects, the effect of temperature, the effect of the
presence of permanent dipoles (and also of free charges) in some
systems—such as those containing water or free ions, and it must also
explore the borderline area between dispersion forces and chemical
bonds. A theory that seeks to do all this is rather ambitious at this stage
of the development of the subject. Our aim in the rest of this book will be
only to indicate possible directions of development of part of such a
theoretical framework.

To end this preliminary chapter we discuss in the concluding section
how the oscillator model of molecules can form the basis of a more
general approach towards evaluation of macroscopic dispersion energy,
than is done in the pairwise summation method.

1.8 Macroscopic dispersion energy from the Drude oscillator model—effect of screening due to induced polarization

We have seen in §1.2 that the dipolar harmonic oscillator model of a
molecule is quite useful for the analysis of dispersion forces when one of
the absorption frequencies is dominant. In this model the interaction
between a pair of molecules can be expressed in terms of the change in
the zero-point energy of the two oscillators due to their mutual induced
dipole-dipole interaction. This method was also developed by London
(1930b). It can be illustrated easily in the example of two identical linear
harmonic oscillators for which the Hamiltonian is

$$H = \frac{m}{2}(\dot{x}_1^2 + \dot{x}_2^2) + \tfrac{1}{2}m\omega_0^2(x_1^2 + x_2^2) - \frac{2e^2}{R^3}x_1x_2, \qquad (1.66)$$

where R is the distance of separation between the centres of the two
oscillators, presumed to be along the x-axis. The last term gives the
interaction energy of the two dipoles when the first oscillator has a dipole
moment $(-ex_1)$ and the second $(-ex_2)$. The potential energy term can be
diagonalized by the change of variables

$$\xi_1 = (x_1 + x_2)/\sqrt{2}, \qquad \xi_2 = (x_1 - x_2)/\sqrt{2} \qquad (1.67)$$

which substitution yields

$$\tfrac{1}{2}m\omega_0^2\left(x_1^2 + x_2^2 - \frac{4e^2}{m\omega_0^2 R^3}x_1 x_2\right)$$

$$= \tfrac{1}{2}m\omega_0^2\left(1 - \frac{2e^2}{m\omega_0^2 R^3}\right)\xi_1^2 + \tfrac{1}{2}m\omega_0^2\left(1 + \frac{2e^2}{m\omega_0^2 R^3}\right)\xi_2^2. \quad (1.68)$$

The kinetic energy part is $(m/2)(\dot\xi_1^2 + \dot\xi_2^2)$, so that we have now two oscillators with effective frequencies

$$\omega^{(\pm)} = \omega_0\sqrt{1 \pm \frac{2e^2}{m\omega_0^2 R^3}}. \quad (1.69)$$

The change in the zero-point energy of the two oscillators, which is the interaction energy, is given by

$$V(R) = \frac{\hbar}{2}(\omega^{(+)} + \omega^{(-)} - 2\omega_0) \approx -\frac{e^4\hbar}{4m^2\omega_0^3 R^6}, \quad (1.70)$$

where we have expanded the square roots of (1.69) up to second order terms in the binomial expansion. A similar analysis can be carried out in the case of three-dimensional oscillators, for which the interaction term is

$$e^2(x_1 x_2 + y_1 y_2 - 2z_1 z_2)/R^3.$$

Langbein (1971a,b) has developed a method for estimating the dispersion interaction between macroscopic bodies based on the oscillator model. We shall give here a method which is equivalent to his approach. Each body can be regarded as an assembly of oscillators. The Hamiltonian of one of the bodies can be written as

$$H = \frac{m}{2}\sum_i (\dot u_i^2 + \omega_i^2 u_i^2) - \frac{e^2}{2}\sum_{i,j}' \mathbf{u}_i \mathcal{T}_{ij} \mathbf{u}_j, \quad (1.71)$$

where

$$\mathcal{T}_{ij} = -\nabla_i \nabla_j \frac{1}{|\mathbf{r}_i - \mathbf{r}_j|}. \quad (1.72)$$

$\mathcal{T}_{ij}$ is the tensor giving the field $\mathbf{r}_i$ due to a dipole source at $\mathbf{r}_j$. The equations of motion of the $\mathbf{u}$'s obtained from (1.71) lead to the secular equation ($\mathcal{I}$ is the unit matrix)

$$D(\omega) \equiv \left|(\omega_i^2 - \omega^2)\mathcal{I} - \frac{e^2}{m}\mathcal{T}\right| = 0. \quad (1.73)$$

This is a $3N \times 3N$ determinant, the i-th diagonal submatrix being $(\omega_i^2 - \omega^2)\mathcal{I}$ and the (ij)-th off-diagonal submatrix being $\mathcal{T}_{ij}$. $\mathcal{T}_{ij}$ for $j = i$ has the

significance of a self-energy term which diverges in this model of point
dipoles. We shall see later that this term can be handled in the case of
molecules of finite size.

In the absence of the dipolar coupling the secular equation is

$$D_0(\omega) = \prod_i |(\omega_i^2 - \omega^2)\mathscr{I}| = 0. \tag{1.74}$$

The change in zero-point energy is then

$$\Delta E = \frac{\hbar}{2} \sum [\{\text{Roots of } D(\omega)\} - \{\text{Roots of } D_0(\omega)\}]$$

$$= \frac{1}{2\pi i} \oint \left(\frac{\hbar\omega}{2}\right) \frac{d}{d\omega} \ln \left(\frac{D(\omega)}{D_0(\omega)}\right) d\omega$$

$$= -\frac{\hbar}{4\pi i} \oint \ln \left|\left(\mathscr{I} - \frac{e^2}{m(\omega_i^2 - \omega^2)} \mathscr{T}_{ij}\right)\right| d\omega, \tag{1.75}$$

where the contour is any contour which encloses the positive real axis in
the ω-plane. The last line is obtained from the previous one by an integration
by parts. Such formulae for additive functions of the roots of a polynomial
are well-known in other branches of physics such as lattice dynamics
(Maradudin et al., 1971), and are based on a standard result of complex
variable theory given, for example, by Titchmarsh (1939). If we choose a
contour illustrated in Fig. 1.7, eqn. (1.75) can be written as

$$\Delta E = \frac{\hbar}{4\pi} \int_{-\infty}^{\infty} \ln \left|\left(\mathscr{I} - \frac{e^2}{m(\omega_i^2 + \xi^2)} \mathscr{T}_{ij}\right)\right| d\xi. \tag{1.76}$$

Here $\xi = i\omega$. We can now use the formula

$$\ln |[\mathscr{I} + \mathscr{S}]| = \sum_{n=1}^{\infty} \frac{(-1)^{n+1}}{n} \text{Tr} [\mathscr{S}^n], \tag{1.77}$$

where Tr denotes the trace, and can expand ΔE as given by (1.76) in a
series in powers of the polarizability

$$\alpha_i(\omega) = \frac{e^2}{m(\omega_i^2 - \omega^2)}. \tag{1.78}$$

Since $\mathscr{T}_{ii}$ does not occur in $D(\omega)$, the result is

$$\Delta E = -\frac{\hbar}{4\pi}\left(\frac{1}{2} \sum_{i \neq j} \int_{-\infty}^{\infty} \alpha_i(i\xi)\mathscr{T}_{ij}\alpha_j(i\xi)\mathscr{T}_{ji} \, d\xi\right.$$

$$\left. + \frac{1}{3} \sum_{i,j,k} \int_{-\infty}^{\infty} \alpha_i(i\xi)\mathscr{T}_{ij}\alpha_j(i\xi)\mathscr{T}_{jk}\alpha_k(i\xi)\mathscr{T}_{ki}\alpha \, d\xi + \cdots\right),$$

$$i \neq j, j \neq k, k \neq i. \tag{1.79}$$

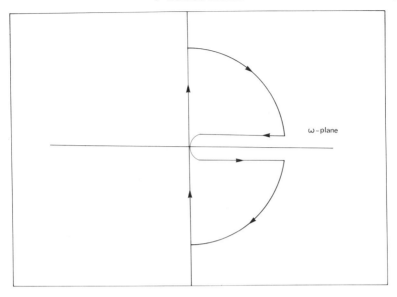

FIG. 1.7. Contour for evaluation of zero-point energy. The original contour encloses the real axis, and can be replaced by an integral from $i\infty$ to $-i\infty$ since the integral around the closed contour shown vanishes, as do those around the semi-circular paths.

If we have two bodies A and B (1.76) can be rearranged to bring out the explicit form of the interaction energy as follows. We can write

$$\left(\mathscr{I} - \frac{e^2}{m(\omega_i^2 + \xi^2)}\,\mathscr{T}_{ij}\right) \equiv [\mathscr{I} - \boldsymbol{\alpha}\mathscr{T}], \qquad (1.80)$$

where

$$\boldsymbol{\alpha} = \begin{pmatrix} \boldsymbol{\alpha}^{(A)} & 0 \\ 0 & \boldsymbol{\alpha}^{(B)} \end{pmatrix} \qquad (1.81)$$

$$\mathscr{T} \equiv \begin{pmatrix} \mathscr{T}^{(A)} & \mathscr{T}^{(C)} \\ \mathscr{T}^{(C)} & \mathscr{T}^{(B)} \end{pmatrix}, \qquad (1.82)$$

where the partitioning implies the separation of the indices i and j into those corresponding to the molecules in the two bodies A and B. For instance, $\boldsymbol{\alpha}^{(A)}$ is a diagonal matrix with 3×3 submatrices of the form $\alpha_i \mathscr{I}$, with

$$\alpha_i(\omega) \equiv \frac{e^2}{m_i(\omega_i^2 - \omega^2)}\, .$$

The submatrices in $\mathscr{T}^{(C)}$ will have the form $\mathscr{T}_{ij}$ with $i \in A$, $j \in B$, and so on. After a little algebra, we can write

$$\ln |[\mathcal{I}-\boldsymbol{\alpha}\mathcal{T}]| = \ln |(\mathcal{I}-\boldsymbol{\alpha}^{(A)}\mathcal{T}^{(A)})| + \ln (\mathcal{I}-\boldsymbol{\alpha}^{(B)}\mathcal{T}^{(B)})|$$
$$+ \ln |(\mathcal{I}-\{(\mathcal{I}-\boldsymbol{\alpha}^{(A)}\mathcal{T}^{(A)})^{-1}\boldsymbol{\alpha}^{(A)}\mathcal{T}^{(C)}\}$$
$$\times \{(\mathcal{I}-\boldsymbol{\alpha}^{(B)}\mathcal{T}^{(B)})^{-1}\boldsymbol{\alpha}^{(B)}\mathcal{T}^{(C)}\})|. \tag{1.83}$$

When this is substituted into (1.76) we note that the first two terms give the dispersion energies of the bodies A and B alone and the last gives the dispersion interaction energy. Thus the interaction energy is

$$E_{AB} \approx -\frac{\hbar}{4\pi}\int_{-\infty}^{\infty} d\xi \operatorname{Tr} (\{(\mathcal{I}-\boldsymbol{\alpha}^{(A)}\mathcal{T}^{(A)})^{-1}\boldsymbol{\alpha}^{(A)}\mathcal{T}^{(C)}$$
$$\times (\mathcal{I}-\boldsymbol{\alpha}^{(B)}\mathcal{T}^{(B)})^{-1}\boldsymbol{\alpha}^{(B)}\tilde{\mathcal{T}}^{(C)}\} + \tfrac{1}{2}\{(\mathcal{I}-\boldsymbol{\alpha}^{(A)}\mathcal{T}^{(A)})^{-1}$$
$$\times \boldsymbol{\alpha}^{(A)}\mathcal{T}^{(C)}(\mathcal{I}-\boldsymbol{\alpha}^{(B)}\mathcal{T}^{(B)})^{-1}\boldsymbol{\alpha}^{(B)}\tilde{\mathcal{T}}^{(C)}\}^2 + \cdots). \tag{1.84}$$

The significance of the matrix $(\mathcal{I}-\boldsymbol{\alpha}^{(A)}\mathcal{T}^{(A)})^{-1}\boldsymbol{\alpha}^{(A)}\mathcal{T}^{(C)}$ can be seen as follows. A typical element with $i \in A$ and $j \in B$ is

$$\{(\mathcal{I}-\boldsymbol{\alpha}^{(A)}\mathcal{T}^{(A)})^{-1}\boldsymbol{\alpha}^{(A)}\mathcal{T}^{(C)}\}_{ij} = \{\boldsymbol{\alpha}^{(A)}\mathcal{T}^{(C)} + \boldsymbol{\alpha}^{(A)}\mathcal{T}^{(A)}\boldsymbol{\alpha}^{(A)}\mathcal{T}^{(C)}$$
$$+ \boldsymbol{\alpha}^{(A)}\mathcal{T}^{(A)}\boldsymbol{\alpha}^{(A)}\mathcal{T}^{(A)}\boldsymbol{\alpha}^{(A)}\mathcal{T}^{(C)} + \cdots\}_{ij}$$
$$= \alpha_i \mathcal{T}^{(C)}_{ij,i\in A,j\in B} + \sum_{\substack{k\in A\\i\in A\\j\in B}} \alpha_i \mathcal{T}^{(A)}_{ik}\alpha_k T^{(C)}_{kj} + \cdots$$
$$= \alpha_i T^{(C)\mathrm{scr}}_{ij} \tag{1.85}$$

where

$$T^{(C)\mathrm{scr}}_{ij} = \mathcal{T}^{(C)}_{ij} + \sum_{k\in A} \mathcal{T}^{(A)}_{ik}\alpha_k \mathcal{T}^{(C)}_{kj} + \cdots. \tag{1.86}$$

The latter is evidently the field at $i \in A$ with a dipole source at $j \in B$ taking into account the induced polarization (screening) on all the molecules in A. A similar interpretation is possible for the elements of $(\mathcal{I}-\boldsymbol{\alpha}^{(B)}\mathcal{T}^{(B)})^{-1}\boldsymbol{\alpha}^{(B)}\mathcal{T}^{(C)}$ which connect the field at a point in B due to a dipole source in A. We can thus write (1.84) as

$$E_{AB} = -\frac{\hbar}{4\pi}\int_{-\infty}^{\infty} d\xi \operatorname{Tr} \left(\sum_{\substack{i,j\\i\in A\\j\in B}} \alpha_i T^{\mathrm{scr}}_{ij}\alpha_j T^{\mathrm{scr}}_{ji} + \frac{1}{2}\sum_{i,j,k}\cdots + \cdots \right). \tag{1.87}$$

This last expression is the basis of Langbein's approach to interactions.

The explicit form of T^{scr}_{ij} can be obtained by a simple artifice. If the bodies A and B are homogeneous, the sums such as $\sum_{k\in A} T_{ik}\alpha_k T_{kj}$ that occur in (1.86) are

$$\sum_{k \in A} T_{ik}\alpha_k T_{ki} = \alpha\rho_A \, \nabla_i \, \nabla_j \int d^3 r_k \, \nabla_k \frac{1}{|\mathbf{r}_i - \mathbf{r}_k|} \nabla_k \frac{1}{|\mathbf{r}_k - \mathbf{r}_j|}. \tag{1.88}$$

The value of the integral depends on the geometrical shape of the body. If the bodies are two planar half-spaces for example, the integral gives us

$$\sum_{k \in A} T_{ik}\alpha_k T_{kj} = \frac{2\pi}{3} \rho_A \alpha^{(A)} \, \nabla_i \, \nabla_j \frac{1}{|\mathbf{r}_i - \mathbf{r}_j|}, \qquad i \in A, \quad j \in B. \tag{1.89}$$

If this is used repeatedly in the series in (1.86) we get

$$T_{ij}^{\text{scr}} = \frac{T_{ij}}{1 + \dfrac{2\pi}{3}\rho_A\alpha^{(A)}}$$

$$T_{ji}^{\text{scr}} = \frac{T_{ji}}{1 + \dfrac{2\pi}{3}\rho_B\alpha^{(B)}} ; \qquad i \in A, \quad j \in B \tag{1.90}$$

If this is now substituted into (1.86) and the Clausius–Mossotti formula used for the dielectric constants of A and B, we would get the Lifshitz (1956) result for the non-retarded interaction energy of two dielectric slabs separated by a distance l

$$\frac{E_{AB}}{\text{surface area}} = -\frac{\hbar}{16\pi^2 l^2} \int_0^\infty d\xi \sum_{n=1}^\infty \frac{1}{n^3}\left(\frac{(\varepsilon_1(i\xi)-1)(\varepsilon_2(i\xi)-1)}{(\varepsilon_1(i\xi)+1)(\varepsilon_2(i\xi)+1)}\right)^n. \tag{1.91}$$

For other geometries the integrand in (1.88) can be expanded in an appropriate set of basis functions which are consistent with the geometrical symmetry of the situation. For instance, for two spheres spherical harmonic expansions can be used to give

$$T_{ij}^{\text{scr}}_{\substack{i \in A \\ j \in B}} = -\nabla_i \, \nabla_j \sum_{m=1}^\infty \frac{1}{\left(1 + \dfrac{4\pi}{3}\rho_1\alpha_1\left(\dfrac{m-1}{2m+1}\right)\right)} \frac{r_i^m}{r_j^{m+1}} Y_m(\theta_{ij}),$$

$$T_{jk}^{\text{scr}} = -\nabla_i \, \nabla_j \sum_{n=1}^\infty \frac{1}{\left(1 + \dfrac{4\pi}{3}\rho_2\alpha_2\left(\dfrac{n-1}{2n+1}\right)\right)} \frac{|\mathbf{r}_j - \mathbf{R}|^n}{|\mathbf{r}_k - \mathbf{R}|^{n+1}} Y_n(\theta_{jk}), \tag{1.92}$$

where the centre of the sphere A is at the origin and that of B is at $\mathbf{R}$. With a fair amount of algebra Langbein (1971b) has shown that this leads to the interaction energy

$$\Delta E_{AB} \approx -\frac{\hbar}{4\pi} \int_{-\infty}^{\infty} d\xi \sum_{m,n=1}^{\infty} \eta_1(m) \left(\frac{b_1}{R}\right)^{2m+1} \eta_2(n)$$
$$\times \left(\frac{b_2}{R}\right)^{2n+1} \binom{2n+2m}{2n}, \quad (1.93)$$

where b_i is the radius of the i-th sphere and

$$\eta_i(n) = \frac{4\pi\rho_i\alpha_i n}{(2n+1) + \frac{4\pi}{3}\rho_i\alpha_i(n-1)}$$

$$= \frac{n(\varepsilon_2-1)}{n\varepsilon_2+n+1}, \quad i=1,2. \quad (1.94)$$

For small separations $l = R - b_1 - b_2 \ll b_1, b_2$ this reduces to

$$|\Delta E_{AB}| \approx \frac{\hbar b_1 b_2}{8\pi l(b_1+b_2)} \int_0^{\infty} d\xi \left(\frac{\varepsilon_1(i\xi)-1}{\varepsilon_1(i\xi)+1}\right)\left(\frac{\varepsilon_2(i\xi)-1}{\varepsilon_2(i\xi)+1}\right) \quad (1.95)$$

provided $\varepsilon_i(i\xi) - 1 \ll 1$. If further the Lorentz relation holds $\varepsilon - 1 = 4\pi\rho\alpha$, and we make the approximation $\varepsilon + 1 \approx 2$, (1.95) reduces further the limiting form of the Hamaker result for spheres

$$\left(E^A \approx -\frac{A}{12}\frac{b_1 b_2}{l(b_1+b_2)}\right).$$

Other limiting cases are given by Langbein (1971b). It should be noted that (1.93) is only the first term in the complete expression (1.87) which is formally exact in the non-retarded limit. The approximate result (1.95) is only useful when $\varepsilon - 1 \ll 1$, i.e., when screening can be ignored. The complete expression is (Langbein, 1971b)

$$\Delta E_{AB} = -\frac{\hbar}{4\pi} \int_{-\infty}^{\infty} d\xi \sum_{j=1}^{\infty} \frac{1}{j} \sum_{m_1,\ldots,m_j=1}^{\infty} \sum_{n_1,n_2,\ldots,n_j=1}^{\infty} C(m,n)$$
$$\times \eta_1(m_1)\left(\frac{b_1}{R}\right)^{2m_1+1} \eta_2(n_1)\left(\frac{b_2}{R}\right)^{2n_1+1} \cdots \eta_1(m_j)\left(\frac{b_1}{R}\right)^{2m_j+1}$$
$$\times \eta_2(n_j)\left(\frac{b_2}{R}\right)^{2m_j+1}, \quad (1.96)$$

where

$$C(m,n) = \sum_{\mu} \binom{m_1+n_1}{m_1+\mu}\binom{n_1+m_2}{n_1+\mu} \cdots \binom{m_j+n_j}{m_j+\mu}\binom{n_j+m_j}{n_j+\mu}, \quad (1.97)$$

the factors are bionomial coefficients, and μ ranges over all integers. While some useful bounds have been obtained by Langbein (1971b), the actual utility of expressions like (1.96) is limited because of computational difficulties. For real situations in colloid science, the real interest is in the magnitude of interaction energies at close separations (b_1/R, $b_2/R \approx 1$). In those circumstances, unless $\varepsilon - 1 \ll 1$, the convergence of series like (1.96) is extremely slow. Thus if one wishes to study interaction energies in general for arbitrary geometries, it is often necessary to resort to other procedures which will be discussed further below.

While the actual application of Langbein's method to specific geometries is then rather involved, it represents nonetheless a distinct important advance over the earlier pairwise summation techniques in view of the fact that it takes into account explicitly the screening effect of the polarization induced on all the molecules in the two bodies. We mention here too that Renne (1971) has developed a formalism which is closely related to the non-retarded method of Langbein and can be reduced to the formulae given by Langbein.

References

Blokhintsev, D. I. (1964). "Quantum Mechanics," D. Reidel Publishing Co., Dordrecht-Holland, §92.

Bouwkamp, C. J. (1947). *Kon. Ned. Akad. Wetenschap.* **50**, 1071.

Bradley, R. S. (1932). *Phil. Mag.* **13**, 853.

Brenner, S. L. and McQuarrie, D. (1973). *Biophys. J.* **13**, 301.

Casimir, H. B. G. and Polder, D. (1948). *Phys. Rev.* **73**, 360.

de Boer, J. H. (1936). *Trans. Farad. Soc.* **32**, 21.

Debye, P. (1912). *Phys. z.* **13**, 97, 295.

Debye, P. (1920). *Phys. z.* **21**, 178; (1921) **22**, 302.

de Rocco, A. G. and Hoover, W. (1960). *Proc. Nat. Acad. Sci. (USA)* **46**, 1057.

Deryaguin, B. V. and Landau, L. D. (1941). *Acta Physica Chem.* **14**, 633; *J. Exp. Theor. Phys.* (1941) **11**, 802; reprinted (1945) **15**, 662.

Eisenschitz, R. and London, F. (1930). *Z. Phys.* **60**, 491.

Elliott, G. F. (1968). *J. theoret. Biol.* **21**, 71.

Ellis, R. (1912). *Z. Phys. Chem.* **80**, 597.

Garstang, R. H. (1966). *Sym. nit. astr. Un.* **26**, 57.
　See also (for transition probabilities)
　1. Glennon, B. M. and Weise, W. L. (1966). U.S. Natl. Bur. Standards Miscellaneous Publication 278. Also "Bibliography of Atomic Transition Publications" (1968) Supplement.
　2. Carliss, C. H. and Bozman, W. R. (1962). U.S. Natl. Bur. Standards Monograph 53.

Gregory, J. (1969). *Adv. Coll. Interface Sci.* **12**, 396.

Hamaker, H. C. (1937). *Physica* **4**, 1058.

Hirschfelder, J. O., Curtiss, C. F. and Bird, R. B. (1954). "Molecular Theory of Gases and Liquids," Wiley, New York.

Israelachvili, J. N. (1973). *J. Chem. Soc. Faraday II* **69**, 1729.

Israelachvili, J. N. (1974). *Quart. Rev. Biophys.* **6**, No. 4, 341.

Israelachvili, J. N. and Tabor, D. (1972a). *Nature (London)* **236**, 106.

Israelachvili, J. N. and Tabor, D. (1972b). *Proc. Roy. Soc. A* **331**, 19.

Israelachvili, J. N. and Tabor, D. (1973). *Progress in Surface and Membrane Science* **7**, 1.

Kallmann, H. and Willstätter, M. (1932). *Naturwiss.* **20**, 952.

Keesom, W. H. (1921). *Phys. z.* **22**, 129; **22**, 643.

Landau, L. D. and Lifshitz, E. M. (1958). "Quantum Mechanics—Non-Relativistic Theory," Pergamon, London.

Langbein, D. (1971a). *J. Chem. Phys. Solids* **32**, 133.

Langbein, D. (1971b). *J. Chem. Phys. Solids* **32**, 1657.

Langmuir, I. (1916). *J. Amer. Chem. Soc.* **38**, 2221.

Langmuir, I. (1917). *J. Amer. Chem. Soc.* **39**, 1848.

Lewis, G. N. (1916). *J. Amer. Chem. Soc.* **38**, 762.

London, F. (1930a). *Z. Phys.* **63**, 245.

London, F. (1930b). *Z. f. Phys. Chem.* B **11**, 222.

Maradudin, A. A., Montroll, E. W., Weiss, G. H. and Ipatova, I. P. (1971). "Theory of Lattice Dynamics in the Harmonic Approximation," Academic Press, New York.

Margenau, H. (1931). *Phys. Rev.* **38**, 747.

Margenau, H. and Kestner, N. R. (1971). "Theory of Intermolecular Forces," Pergamon, Oxford (2nd Edition).

Maxwell, J. C. (1875). "Capillary Action"—Article in Encyclopaedia Britannica, 9th edition, updated by Lord Rayleigh in 10th ed. See also Lord Rayleigh, this list.

Mitchell, D. J. and Ninham, B. W. (1973). *J. Chem. Phys.* **59**, 1246.

Onsager, L. (1942). *Phys. Rev.* **62**, 558.

Onsager, L. (1944). *Ann. N.Y. Acad. Sci.* **15**, 627.

Parsegian, V. A. (1973). *Ann. Rev. Biophys. and Bioeng.* **2**, 221.

Pauling, L. and Beach, J. Y. (1935). *Phys. Rev.* **47**, 686.

Poisson, S. D. (1831). See Maxwell (1875) and Lord Rayleigh (1890) for remarks on Poisson's theory.

Rayleigh, Lord (1890). *Phil. Mag.* **30**, 285, 456. [See also Scientific Papers (1964) v. 3, articles 176 and 186. Dover, New York.]

Renne, M. J. (1971). *Physica* **56**, 125.

Reinganum, M. (1912). *Ann. d. Phys.* **38**, 649.

Salem, L. (1962). *J. Chem. Phys.* **37**, 2100.

Schiff, L. I. (1968). "Quantum Mechanics," 3rd ed. McGraw-Hill, New York, §32.

Slater, J. C. and Kirkwood, J. (1931). *Phys. Rev.* **37**, 682.

Sparnaay, M. J. (1959). *Recueil de Travaux Chemique des Pays-Bas* **78**, 680.

Thomson, J. J. (1914). *Phil. Mag.* **27**, 757.

Titchmarsh, E. C. (1939). "The Theory of Functions," 2nd ed. ch. 3. Oxford Univ. Press.

Tomlinson, G. A. (1928). *Phil. Mag.* **6**, 695.

Tomlinson, G. A. (1930). *Phil. Mag.* **10**, 541.

Triezenberg, D. G. and Zwanzig, R. (1972). *Phys. Rev. Letts.* **28**, 1183.

van der Waals, J. H. (1873). Ph.D. Thesis, University of Leiden.

van der Waals, J. D. (Jr) (1909). *Amst. Acad. Proc.* pp. 132, 315.

van Urk, A. Th. A. (1937). *Physica* **4**, 1025.

Verwey, E. J. and Overbeek, J. Th. G. (1948). "Theory of the Stability of Lyophobic Colloids," Elsevier, New York.

Vold, M. J. (1954). *J. Coll. Interface Sci.* **9**, 451.

Vold, M. J. (1957). *Proc. Ind. Acad. Sci.* A **46**, 152.

Wang, S. C. (1927). *Phys. z.* **28**, 663.

Widom, B. (1965). *J. Chem. Phys.* **43**, 3892; **43**, 3898.

Young, T. (1805). "An Essay on the Cohesion of Fluids," *Phil. Trans.* **I**, 65. [See also works of Dr. Young, vols. I and II (1855) (George Peacock, ed.). John Murray, London.]

Chapter 2

Dispersion Forces from the Field Point of View

2.1 General remarks

In Chapter 1 we have seen that the origin of the dispersion force between two molecules is linked to a process which can be described as the induction of polarization on one due to the instantaneous polarization field of the other, and the value of the dispersion interaction energy is the expectation value of the corresponding interaction term in the Hamiltonian. Since the interaction occurs through the electromagnetic field, it stands to reason that an alternative viewpoint could be developed, according to which the dispersion interaction of a pair of molecules could be considered to be due to the effect of the pair on the energy of the electromagnetic field.

Historically, this approach was developed in a series of papers by Casimir (1948, 1949) and by Casimir and Polder (1946, 1948). An important result obtained by Casimir and Polder using quantum electrodynamics was that the dispersion interaction energy between a pair of molecules at a distance larger than the wavelength of the radiation due to dipolar transitions in them falls off as $(1/R^7)$, according to the formula

$$E(R) = -\frac{23}{4\pi} \hbar c \frac{\alpha_1(0)\alpha_2(0)}{R^7}. \tag{2.1}$$

The quantum electrodynamic approach developed by Casimir and Polder (1948) was also formulated by Casimir in semi-classical terms, in which the interaction energy can be defined as the change in the zero-point energy of the electromagnetic field modes (obtained by solving Maxwell's equations) when the latter are perturbed by the molecules through coupling of the field with the polarization currents induced on the molecules. This kind of semi-classical approach has attracted renewed interest lately in the study of problems involving interaction of radiation with matter (Scully and Sargent, 1972) with the development of lasers. We shall restrict our considerations here within the semi-classical approach, in view of the considerable simplification in the mathematical aspects of the framework over the approach based on quantum electrodynamics. A simple illustration of the method is in the derivation of the force between two perfectly conducting metallic plates by Casimir (1948). Consider two perfectly conducting plates separated by a distance l along the z-direction, with the (x, y)-axes lying on one of them. The modes of the electric field can be written as

$$\mathbf{E}(k_1, k_2, n) = \mathbf{E}_0 e^{i(k_1 x + k_2 y)} \sin \frac{n\pi z}{l}, \tag{2.2}$$

where k_1 and k_2 are the wave numbers of propagating waves along the x- and y-directions respectively, and $(n-1)$ is the number of standing wave modes along the z-direction. The corresponding frequency is

$$\omega(k_1, k_2, n) = c\left(k_1^2 + k_2^2 + \frac{n^2\pi^2}{l^2}\right)^{1/2}. \tag{2.3}$$

In addition, there will be one mode for $n = 0$. For other integral values of n there will be two modes corresponding to two polarizations. The interaction energy per unit area between the plates can be defined as,

$$
\begin{aligned}
E(l) = {} & \frac{\hbar}{2} \sum_{k_1, k_2\, n} \left[\omega(k_1, k_2, n) - \lim_{l \to \infty} \omega(k_1, k_2, n)\right] \\
= {} & \frac{\hbar c}{2} \frac{2}{(2\pi)^2} \iint_{-\infty}^{\infty} dk_1\, dk_2 \left(\sum_{n=1}^{\infty} \sqrt{k_1^2 + k_2^2 + n^2\pi^2/l^2}\right. \\
& \left. + \tfrac{1}{2}\sqrt{k_1^2 + k_2^2} - \int_0^{\infty} \sqrt{k_1^2 + k_2^2 + n^2\pi^2/l^2}\, dn\right) \\
= {} & \frac{\hbar c}{(2\pi)} \int_0^{\infty} \kappa\, d\kappa \left(\sum_{n=0}^{\infty\,\prime} \sqrt{\kappa^2 + n^2\pi^2/l^2} - \int_0^{\infty} \sqrt{\kappa^2 + n^2\pi^2/l^2}\, dn\right),
\end{aligned}
\tag{2.4}
$$

where the prime is meant to indicate that in the sum the term for $n = 0$ is to be multiplied by $\tfrac{1}{2}$. The integral over n is an estimate of the sum for large l.

The individual integrals diverge but their difference does not. To extract a meaningful result we can introduce a convergence factor $e^{-(\kappa^2+n^2\pi^2/l^2)\delta^2}$ with $\delta \to 0$ ultimately. Then we can define a function

$$S(\delta, n) = \int_0^\infty \kappa\, d\kappa\, e^{-(\kappa^2+n^2\pi^2/l^2)\delta^2}\sqrt{\kappa^2 + n^2\pi^2/l^2}$$

$$= \frac{1}{2}\int_{n^2\pi^2/l^2}^\infty du\sqrt{u}e^{-u\delta^2},$$

in terms of which the above sum can be expressed. We can now use the Euler–Maclaurin formula to simplify the sum in (2.4) as,

$$\sum_{n=0}^{\infty}{}' S(\delta, n) - \int_0^\infty S(\delta, n)\, dn = \tfrac{1}{6}[S'(\delta, n = \infty) - S'(\delta, n = 0)]$$

$$-\frac{1}{30\times 24}[S'''(\delta, n = \infty) - S'''(\delta, n = 0)] + \cdots . \quad (2.5)$$

The derivatives of $S(\delta, n)$ (which are with respect to n) are

$$S'(\delta, n) = n^2\left(\frac{\pi}{l}\right)^3 e^{-n^2\pi^2\delta^2/l^2}$$

$$S'''(\delta, n) = \left(\frac{\pi}{l}\right)^3\left\{2 + 8n^2\left(\frac{\pi^2\delta^2}{l^2}\right) + n^2\left(\frac{2\pi^2\delta^2}{l^2} - 4n^2\frac{\pi^4\delta^4}{l^4}\right)\right\}e^{-n^2\pi^2\delta^2/l^2}$$

The limits for $n = \infty$ and $n = 0$ are taken trivially and we get for $\delta \to 0$,

$$E(l) \cong -\frac{\hbar c}{(2\pi)}\frac{2\pi^3}{l^3}\frac{1}{30\times 24} = -\frac{\hbar c\pi^2}{720 l^3}. \quad (2.6)$$

This would lead to an attractive force per unit area between the plates, whose value is

$$F = -\frac{\partial E(l)}{\partial l} = -\frac{\hbar c\pi^2}{240}\frac{1}{l^4}. \quad (2.7)$$

This force arises from the change in the zero-point energy per unit volume of the electromagnetic field from the free space value to what it is when the field is confined between the two plates separated by the distance l. If the distance l is measured in microns, the numerical value of F is,

$$|F| = \frac{0\cdot 013}{l^4}\ \text{dynes/cm}^2. \quad (2.8)$$

Since no metal in nature is an ideal conductor, this result may be expected to be valid as long as l is larger than the skin-depth, i.e., the penetration depth of electromagnetic waves, which for most metals is of the order of 0.1 μ. We shall later show that a similar $(1/l^4)$ force arises between two dielectric plates when l exceeds the characteristic absorption wavelengths of the media (Lifshitz, 1954, 1955).

This result is very different from the $(1/l^3)$ law that was obtained in §1.4 for the force between slabs by the pairwise summation of London $(1/R^6)$ force between the constituent molecules in the slabs. The dispersion force between a pair of molecules in this retarded region, where the finite velocity of propagation of electromagnetic interactions begins to be felt, has a different character from the London $(1/R^6)$ force and will be investigated in the next section. The force between metals can also be derived by a different approach which emphasizes the surface current fluctuations as the source of radiation (Mitchell, Ninham and Richmond, 1972).

2.2   Force between molecules in the retarded region

The effect of the electromagnetic field on a molecule embedded in it is mainly in the form of induction of polarization currents on it. If a molecule is subjected to a variable electric field

$$\mathbf{E}(t) = \mathbf{E}_0(\omega) \cos \omega t, \tag{2.9}$$

the polarization induced on it is given by

$$\mathbf{P}(t) = \mathbf{P}(\omega) \cos \omega t = [\boldsymbol{\alpha}(\omega)\mathbf{E}_0(\omega)] \cos \omega t, \tag{2.10}$$

where the tensor $\boldsymbol{\alpha}(\omega)$ has components

$$\alpha_{xy}(\omega) = -\frac{1}{\hbar} \sum_n \left(\frac{(\mathbf{D}_{ln})_y(\mathbf{D}_{nl})_x}{\omega_{nl} - \omega} + \frac{(\mathbf{D}_{nl})_y(\mathbf{D}_{ln})_x}{\omega_{nl} + \omega} \right), \tag{2.11}$$

where $\mathbf{D} = -\sum e_j \mathbf{r}_j$, the dipole moment operator of the molecule. The form of $\boldsymbol{\alpha}(\omega)$ is derived in many standard texts on quantum mechanics (see for instance, Blokhintsev (1964)). We shall see later in Chapter 4 that a more appropriate description of induced polarization would have to take the size of the molecule into account. But, for the present, we shall assume that the molecules are essentially point-particles, so that in the presence of an electric field $\mathbf{E}(\mathbf{r}, t)$ the induced polarization density can be written as

$$\mathbf{p}(\mathbf{r}, t) = \left(\int_{-\infty}^{\infty} \boldsymbol{\alpha}(t - t')\mathbf{E}(\mathbf{R}, t') \, dt' \right) \delta(\mathbf{r} - \mathbf{R}), \tag{2.12}$$

where $\mathbf{R}$ is the position coordinate of the molecule. Here $\boldsymbol{\alpha}(t)$ is the Fourier transform of $\boldsymbol{\alpha}(\omega)$ defined by (2.11),

$$\boldsymbol{\alpha}(t) = \frac{1}{2\pi} \int_{-\infty}^{\infty} d\omega e^{i\omega t} \boldsymbol{\alpha}(\omega), \tag{2.13}$$

and $E(r, t)$ can be written in the form

$$E(r, t) = \frac{1}{2\pi} \int_{-\infty}^{\infty} d\omega E(r, \omega) e^{i\omega t}. \tag{2.14}$$

Eqn. (2.12) follows from the fact that the polarization developed on the molecule is the sum of that due to each Fourier frequency component of the electric field. In general, $\alpha(\omega)$ can be defined with a small imaginary part in ω, such that its Fourier transform $\alpha(t)$ vanishes for $t < 0$, so as to satisfy the principle of causality. Thus, in eqn. (2.12) effectively the upper limit of integration is t, and its Fourier transform can be written as,

$$p(r, \omega) = [\alpha(\omega)E(R, \omega)]\, \delta(r - R), \tag{2.15}$$

where

$$p(r, \omega) = \int_{-\infty}^{\infty} e^{-i\omega t} p(r, t)\, dt. \tag{2.16}$$

To consider the effect of the coupling between a pair of molecules at R_1 and R_2 with polarizabilities $\alpha_1(\omega)$ and $\alpha_2(\omega)$, and the electromagnetic field, we note that the polarization current induced on the molecules is

$$j(r, t) = \frac{\partial}{\partial t} p(r, t), \tag{2.17}$$

and the corresponding Fourier transforms, defined as in eqn. (2.16), satisfy the relation

$$j(r, \omega) = i\omega p(r, \omega)$$
$$= i\omega[\alpha_1(\omega)E(R_1, \omega)\, \delta(r - R_1) + \alpha_2(\omega)E(R_2, \omega)\, \delta(r - R_2)]. \tag{2.18}$$

The electromagnetic field equations can now be written down with the source terms corresponding to the current $j(r, t)$. It is convenient to use the equations for the vector potential A and scalar potential ϕ in the Lorentz gauge for this analysis. The necessary equations are,

$$\left(\nabla^2 - \frac{1}{c^2}\frac{\partial^2}{\partial t^2}\right)A(r, t) = -\frac{4\pi}{c}j(r, t) \tag{2.19}$$

and

$$\nabla \cdot A + \frac{1}{c}\frac{\partial \phi}{\partial t} = 0. \tag{2.20}$$

The electric field $E(r, t)$ is given by

$$E(r, t) = -\frac{1}{c}\frac{\partial A}{\partial t} - \nabla\phi. \tag{2.21}$$

The time Fourier transform of the above three equations are,

$$\left(\nabla^2 + \frac{\omega^2}{c^2}\right)\mathbf{A}(\mathbf{r}, \omega) = -\frac{4\pi}{c}\mathbf{j}(\mathbf{r}, \omega) \tag{2.22}$$

$$\nabla \cdot \mathbf{A}(\mathbf{r}, \omega) + \frac{i\omega}{c}\phi(\mathbf{r}, \omega) = 0 \tag{2.23}$$

$$\mathbf{E}(\mathbf{r}, \omega) = -\frac{i\omega}{c}\mathbf{A}(\mathbf{r}, \omega) - \nabla\phi(\mathbf{r}, \omega)$$

$$= -\frac{ic}{\omega}\left(\frac{\omega^2}{c^2}\mathbf{A} + \nabla(\nabla \cdot \mathbf{A})\right). \tag{2.24}$$

Using eqns. (2.18) and (2.24) in (2.22) we get

$$\left(\nabla^2 + \frac{\omega^2}{c^2}\right)\mathbf{A}(\mathbf{r}, \omega) = -4\pi[\alpha_1(\omega)\,\delta(\mathbf{r} - \mathbf{R}_1) + \alpha_2(\omega)\,\delta(\mathbf{r} - \mathbf{R}_2)]$$

$$\times\left(\frac{\omega^2}{c^2}\right)\mathbf{A}(\mathbf{r}, \omega) + \nabla\{\nabla \cdot \mathbf{A}(\mathbf{r}, \omega)\}]. \tag{2.25}$$

Since the operator $[(\omega^2/c^2)\mathscr{I} + \nabla(\nabla\cdot)]$ commutes with $[\nabla^2 + (\omega^2/c^2)]$, the equation that $\mathbf{E}(\mathbf{r}, \omega)$ satisfies, becomes

$$\left(\nabla^2 + \frac{\omega^2}{c^2}\right)\mathbf{E}(\mathbf{r}, \omega) = -4\pi\left\{\alpha_1(\omega)\left[\left(\frac{\omega^2}{c^2}\right)\mathbf{E}(\mathbf{R}_1, \omega)\,\delta(\mathbf{r} - \mathbf{R}_1)\right.\right.$$

$$\left. + \mathbf{\Omega}(\mathbf{r}, \mathbf{R}_1)\mathbf{E}(\mathbf{R}_1, \omega)\right] + \alpha_2(\omega)\left[\left(\frac{\omega^2}{c^2}\right)\mathbf{E}(\mathbf{R}_2, \omega)\,\delta(\mathbf{r} - \mathbf{R}_2)\right.$$

$$\left.\left. + \mathbf{\Omega}(\mathbf{r}, \mathbf{R}_2)\mathbf{E}(\mathbf{R}_2, \omega)\right]\right\}, \tag{2.26}$$

where $\mathbf{\Omega}(\mathbf{r}, \mathbf{r}')$ is the dyadic $\nabla_r\nabla_r\,\delta(\mathbf{r} - \mathbf{r}')$, explicitly

$$\mathbf{\Omega}(\mathbf{r}, \mathbf{r}') \equiv \begin{pmatrix} \dfrac{\partial^2}{\partial x^2} & \dfrac{\partial^2}{\partial x\,\partial y} & \dfrac{\partial^2}{\partial x\,\partial z} \\[2mm] \dfrac{\partial^2}{\partial y\,\partial x} & \dfrac{\partial^2}{\partial y^2} & \dfrac{\partial^2}{\partial y\,\partial z} \\[2mm] \dfrac{\partial^2}{\partial z\,\partial x} & \dfrac{\partial^2}{\partial z\,\partial y} & \dfrac{\partial^2}{\partial z^2} \end{pmatrix}\delta(\mathbf{r} - \mathbf{r}'). \tag{2.27}$$

Equation (2.26) can be written as an integral equation,

$$\mathbf{E}(\mathbf{r}, \omega) = -4\pi[\alpha_1(\omega)\mathscr{G}(\mathbf{r}, \mathbf{R}_1; \omega)\mathbf{E}(\mathbf{R}_1, \omega) + \alpha_2(\omega)\mathscr{G}(\mathbf{r}, \mathbf{R}_2; \omega)\mathbf{E}(\mathbf{R}_2, \omega)], \tag{2.28}$$

where

$$\mathscr{G}(\mathbf{r}, \mathbf{r}'; \omega) = \frac{\omega^2}{c^2} \mathscr{G}(\mathbf{r} - \mathbf{r}'; \omega) + \nabla_r \nabla_r \mathscr{G}(\mathbf{r} - \mathbf{r}'; \omega) \tag{2.29}$$

where

$$\mathscr{G}(\mathbf{r} - \mathbf{r}'; \omega) = \mathscr{I} \frac{1}{(2\pi)^3} \int \frac{e^{i\mathbf{k}.(\mathbf{r} - \mathbf{r}')} \, d^3 k}{\omega^2/c^2 - k^2} \,. \tag{2.30}$$

$\mathscr{G}(\mathbf{r} - \mathbf{r}'; \omega)$ is the dyadic Green function of the free field equation for $\mathbf{E}$.
From eqn. (2.28) we get two equations for $\mathbf{E}(\mathbf{R}_1, \omega)$ and $\mathbf{E}(\mathbf{R}_2, \omega)$,

$$\mathbf{E}(\mathbf{R}_1, \omega) = -4\pi[\boldsymbol{\alpha}_1(\omega)\mathscr{G}(\mathbf{R}_1, \mathbf{R}_1; \omega)\mathbf{E}(\mathbf{R}_1, \omega)$$
$$+ \boldsymbol{\alpha}_2(\omega)\mathscr{G}(\mathbf{R}_1, \mathbf{R}_2; \omega)\mathbf{E}(\mathbf{R}_2, \omega)]$$
$$\mathbf{E}(\mathbf{R}_2, \omega) = -4\pi[\boldsymbol{\alpha}_1(\omega)\mathscr{G}(\mathbf{R}_2, \mathbf{R}_1; \omega)\mathbf{E}(\mathbf{R}_1, \omega)$$
$$+ \boldsymbol{\alpha}_2(\omega)\mathscr{G}(\mathbf{R}_2, \mathbf{R}_2; \omega)\mathbf{E}(\mathbf{R}_2, \omega)],$$

and for solutions to exist the secular determinant formed out of the coefficients must vanish. Thus, we get the secular equation

$$\begin{vmatrix} I + 4\pi\boldsymbol{\alpha}_1(\omega)\mathscr{G}(\mathbf{R}_1, \mathbf{R}_1; \omega) & 4\pi\boldsymbol{\alpha}_2(\omega)\mathscr{G}(\mathbf{R}_1, \mathbf{R}_2; \omega) \\ 4\pi\boldsymbol{\alpha}_1(\omega)\mathscr{G}(\mathbf{R}_2, \mathbf{R}_1; \omega) & 1 + 4\pi\boldsymbol{\alpha}_2(\omega)\mathscr{G}(\mathbf{R}_2, \mathbf{R}_2; \omega) \end{vmatrix} = 0. \tag{2.31}$$

The roots of this give the perturbed frequencies of the modes of the electromagnetic field, the perturbation being due to the two molecules. The way this secular equation has been arrived at implies that the above determinant is the ratio of $D_{12}(\omega)$ and $D_0(\omega)$, the former being the secular determinant of the field equation with the two molecules present and the latter being that for the free field. By reasoning similar to that leading to eqn. (2.31) we can show that if $D_1(\omega)$ and $D_2(\omega)$ respectively are the secular determinants with the first and second molecules coupled to the field, then

$$\frac{D_j(\omega)}{D_0(\omega)} = |\mathscr{I} + 4\pi\boldsymbol{\alpha}_j(\omega)\mathscr{G}(\mathbf{R}_j, \mathbf{R}_j; \omega)|; \qquad j = 1, 2. \tag{2.32}$$

The change in the zero-point energy of the field is,

$$\Delta E_{12} = \frac{\hbar}{2}\Big[\sum\{\text{Roots of } D_{12}(\omega) - \text{Roots of } D_0(\omega)\}$$
$$- \sum\{\text{Roots of } D_1(\omega) - \text{Roots of } D_0(\omega)\}$$
$$- \sum\{\text{Roots of } D_2(\omega) - \text{Roots of } D_0(\omega)\}\Big].$$

If we use the contour integral representation used in §1.8, the above can be written as

$$\Delta E(R_{12})$$

$$= \frac{1}{2\pi i} \oint \left(\frac{\hbar\omega}{2}\right) d\omega \frac{d}{d\omega} \ln \left| \frac{D_{12}(\omega)/D_0(\omega)}{\{D_1(\omega)/D_0(\omega)\}\{D_2(\omega)/D_0(\omega)\}} \right|$$

$$= -\frac{\hbar}{4\pi i} \oint d\omega \ln \frac{\left| \begin{matrix} \mathcal{I} + 4\pi\boldsymbol{\alpha}_1(\omega)\mathcal{G}(\mathbf{R}_1, \mathbf{R}_1; \omega) & 4\pi\boldsymbol{\alpha}_2(\omega)\mathcal{G}(\mathbf{R}_1, \mathbf{R}_2; \omega) \\ 4\pi\boldsymbol{\alpha}_1(\omega)\mathcal{G}(\mathbf{R}_2, \mathbf{R}_1; \omega) & \mathcal{I} + 4\pi\boldsymbol{\alpha}_2(\omega)\mathcal{G}(\mathbf{R}_2, \mathbf{R}_2; \omega) \end{matrix} \right|}{\left| \mathcal{I} + 4\pi\boldsymbol{\alpha}_1(\omega)\mathcal{G}(\mathbf{R}_1, \mathbf{R}_1; \omega) \right| \left| \mathcal{I} + 4\pi\boldsymbol{\alpha}_2(\omega)\mathcal{G}(\mathbf{R}_2, \mathbf{R}_2; \omega) \right|}.$$

$$(2.33)$$

At this point, it must be stated that $\mathcal{G}(\mathbf{R}_1, \mathbf{R}_1; \omega)$ is an undefined quantity for a point-molecule, because the components of the tensor $\mathcal{G}(\mathbf{r}, \mathbf{r}'; \omega)$ diverge for $\mathbf{r} \rightarrow \mathbf{r}'$. But actual molecules are not geometrical points, and it will be shown in Chapter 4 that the finite size of the molecules would be responsible for keeping $\mathcal{G}(\mathbf{R}_j, \mathbf{R}_j; \omega)$ finite. In any case, since we are interested only in the interaction energy, we can use the formula of eqn. (1.84) to get,

$$V(R_{12}) \cong -\frac{4\pi\hbar}{i} \oint d\omega \, \mathrm{Tr} \{\boldsymbol{\alpha}_1(\omega)\mathcal{G}(\mathbf{R}_1, \mathbf{R}_2; \omega)\boldsymbol{\alpha}_2(\omega)\mathcal{G}(\mathbf{R}_2, \mathbf{R}_1; \omega)\}$$

$$= -8\pi\hbar \int_0^\infty d\xi \, \mathrm{Tr} \{\boldsymbol{\alpha}_1(i\xi)\mathcal{G}(\mathbf{R}_1, \mathbf{R}_2; i\xi)$$

$$\times \boldsymbol{\alpha}_2(i\xi)\mathcal{G}(\mathbf{R}_2, \mathbf{R}_1; i\xi)\}, \qquad (2.34)$$

where we have taken the contour along the imaginary axis, as in eqn. (1.83), and retained terms only up to the product of the two polarizabilities.

If we assume that $\boldsymbol{\alpha}_j(\omega)$ is isotropic, i.e.,

$$\boldsymbol{\alpha}_j(\omega) = \mathcal{I}\alpha_j(\omega), \qquad (2.35)$$

and with a suitable choice of axes (z along $\mathbf{R}_{12}$) $\mathcal{G}(\mathbf{R}_1, \mathbf{R}_2; i\xi)$ can be expressed in a diagonal form

$$\mathcal{G}(\mathbf{R}_1, \mathbf{R}_2; i\xi)$$

$$= \frac{1}{(2\pi)^3} \int \frac{e^{i\mathbf{k}\cdot(\mathbf{R}_1 - \mathbf{R}_2)}[\xi^2/c^2\mathcal{I} + (\mathbf{kk})] \, d^3k}{\xi^2/c^2 + k^2}$$

$$= \frac{e^{-\xi R_{12}/c}}{4\pi R_{21}} \begin{pmatrix} \dfrac{\xi^2}{c^2} + \dfrac{\xi}{cR_{12}} + \dfrac{1}{R_{12}^2} & 0 & 0 \\[2ex] 0 & \dfrac{\xi^2}{c^2} + \dfrac{\xi}{cR_{12}} + \dfrac{1}{R_{12}^2} & 0 \\[2ex] 0 & 0 & -\dfrac{2\xi}{cR_{12}} - \dfrac{2}{R_{12}^2} \end{pmatrix} \qquad (2.36)$$

We can obtain from eqn. (2.34) the equation

$$
\begin{aligned}
V(R_{12}) = & -8\pi\hbar \int_0^\infty d\xi\, \alpha_1(i\xi)\alpha_2(i\xi) \frac{e^{-2\xi R_{12}/c}}{(4\pi R_{12})^2} \\
& \times \left\{ 2\left(\frac{\xi^2}{c^2} + \frac{\xi}{cR_{12}} + \frac{1}{R_{12}^2}\right)^2 + 4\left(\frac{\xi}{cR_{12}} + \frac{1}{R_{12}^2}\right)^2 \right\} \\
= & -\frac{\hbar}{\pi R_{12}^2} \int_0^\infty d\xi\, \alpha_1(i\xi)\alpha_2(i\xi) e^{-2\xi R_{12}/c} \\
& \times \left(\frac{\xi^4}{c^4} + \frac{2\xi^3}{c^3 R_{12}} + \frac{5\xi^2}{c^2 R_{12}^2} + \frac{6\xi}{cR_{12}^3} + \frac{3}{R_{12}^4}\right).
\end{aligned}
\tag{2.37}
$$

The full expression for the dispersion energy between two neutral atoms must also include the contribution from magnetic dipole fluctuations. These too are non-retarded at short range and retarded at long range (Mavroyannis and Stephen, 1962; Feinberg and Sucher, 1970). To proceed further we need information on the detailed nature of $\alpha_j(i\xi)$, $j = 1$, 2. If $\alpha_j(\omega)$ has a form like that chosen in eqn. (1.19), the major part of the contribution of $\alpha_j(i\xi)$ to the integral in (2.37) would come from the neighbourhood of $\xi \lesssim \xi_j (= i\omega_j)$, $j = 1$, 2. If $R_{12} \ll c/\omega_j$, i.e., if the distance of separation is less than the wavelengths of the radiation associated with the transitions of the molecules, then the exponential term can be equated to unity, and the largest term in the parenthesis will be $3/R_{12}^4$. We then get the non-retarded London result,

$$
V(R_{12}) \cong -\frac{3\hbar}{\pi R_{12}^6} \int_0^\infty d\xi\, \alpha_1(i\xi)\alpha_2(i\xi),
\tag{2.38}
$$

which was obtained in eqn. (1.18) by a different argument.

On the other hand, if $R_{12} \gg c/\omega_j$, the main contribution would come from the neighbourhood of $\xi \to 0$. We can then make a Taylor expansion of $\alpha_j(i\xi)$ and the leading term is

$$
\begin{aligned}
V(R_{12}) \cong & -\frac{\hbar}{\pi R_{12}^2} \alpha_1(0)\alpha_2(0) \int_0^\infty d\xi\, e^{-2\xi R_{12}/c} \left(\frac{\xi^4}{c^4} + \frac{2\xi^3}{c^3 R_{12}}\right. \\
& \left. + \frac{5\xi^2}{c^2 R_{12}^2} + \frac{6\xi}{cR_{12}^3} + \frac{3}{R_{12}^4}\right) \\
= & -\frac{\hbar c \alpha_1(0)\alpha_2(0)}{\pi R_{12}^7} \int_0^\infty dx\, e^{-2x}(x^4 + 2x^3 + 5x^2 + 6x + 3) \\
= & -\frac{23\hbar c \alpha_1(0)\alpha_2(0)}{4\pi R_{12}^7}.
\end{aligned}
\tag{2.39}
$$

This is the result of Casimir and Polder (1948) valid in the retarded region. We thus see that consideration of the effect of the two molecules on the zero-point energy of the electromagnetic field leads automatically to the London $(1/R^6)$ interaction in the non-retarded region and the Casimir-Polder $(1/R^7)$ interaction in the retarded region. The London limit is obtained also if $c \to \infty$ in eqn. (2.37).

2.3 Many-body effects

This approach can be extended readily to establish the nature of dispersion forces when there are more than two molecules. If we have n molecules with polarizabilities $\boldsymbol{\alpha}_1(\omega)$, $\boldsymbol{\alpha}_2(\omega)$, ..., $\boldsymbol{\alpha}_n(\omega)$, the equation corresponding to (2.33) becomes

$$\Delta E(n) = -\frac{\hbar}{4\pi i}\oint d\omega \ln \frac{\{D_n(\omega)/D_0(\omega)\}}{\prod\limits_{j=1}^{n}\{D_j(\omega)/D_0(\omega)\}}, \qquad (2.40)$$

where

$\{D_n(\omega)/D_0(\omega)\} =$

$$\begin{pmatrix} \mathscr{I}+4\pi\boldsymbol{\alpha}_1(\omega)\mathscr{G}(\mathbf{R}_1,\mathbf{R}_1;\omega) & 4\pi\boldsymbol{\alpha}_2(\omega)\mathscr{G}(\mathbf{R}_1,\mathbf{R}_2;\omega) & \cdots & 4\pi\boldsymbol{\alpha}_n(\omega)\mathscr{G}(\mathbf{R}_1,\mathbf{R}_n;\omega) \\ 4\pi\boldsymbol{\alpha}_1(\omega)\mathscr{G}(\mathbf{R}_2,\mathbf{R}_1;\omega) & \mathscr{I}+4\pi\boldsymbol{\alpha}_2(\omega)\mathscr{G}(\mathbf{R}_2,\mathbf{R}_2;\omega) & \cdots & \cdots \\ \cdot & \cdot & \cdots & \cdot \\ \cdot & \cdot & \cdots & \cdot \\ \cdot & \cdot & \cdots & \cdot \\ 4\pi\boldsymbol{\alpha}_1(\omega)\mathscr{G}(\mathbf{R}_n,\mathbf{R}_1;\omega) & \cdots & \cdots & \mathscr{I}+4\pi\boldsymbol{\alpha}_n(\omega)\mathscr{G}(\mathbf{R}_n,\mathbf{R}_n;\omega) \end{pmatrix}.$$

$$(2.41)$$

Using the formula of eqn. (1.84) we can write

$$\ln\left(\frac{\{D_n(\omega)/D_0(\omega)\}}{\prod\limits_{j=1}^{n}D_j(\omega)/D_0(\omega)}\right)$$

$$= \mathrm{Tr}\left(-\frac{1}{2}\sum_{i,j}'(4\pi)^2\boldsymbol{\alpha}_i\mathscr{G}(\mathbf{R}_i,\mathbf{R}_j)\boldsymbol{\alpha}_j\mathscr{G}(\mathbf{R}_j,\mathbf{R}_i)\right.$$

$$\left.+\frac{1}{3}\sum_{i,j,k}'(4\pi)^3\boldsymbol{\alpha}_i\mathscr{G}(\mathbf{R}_i,\mathbf{R}_j)\boldsymbol{\alpha}_j\mathscr{G}(\mathbf{R}_j,\mathbf{R}_k)\boldsymbol{\alpha}_k\mathscr{G}(\mathbf{R}_k,\mathbf{R}_i)+\cdots\right). \quad (2.42)$$

The prime over the summation implies that no two summation indices equal each other. We can thus write

$$\Delta E(n) = -\frac{1}{2}\sum_{i,j}'V_2(ij)+\frac{1}{3}\sum_{i,j,k}'V_3(ijk)$$

$$-\frac{1}{4}\sum_{i,j,k,l}'V_4(ijkl)+\cdots, \qquad (2.43)$$

where $V_2(ij)$ is the two-particle dispersion interaction evaluated in (2.37), $V_3(ijk)$ is a three-particle interaction between the molecules with indices i, j, k, and so on. The three-body term is

$$V_3(ijk) = 32\pi^2\hbar \int_0^\infty d\xi\, \text{Tr}\,\{\boldsymbol{\alpha}_i \mathscr{G}(\mathbf{R}_i, \mathbf{R}_j)\boldsymbol{\alpha}_j \mathscr{G}(\mathbf{R}_j, \mathbf{R}_k)\boldsymbol{\alpha}_k \mathscr{G}(\mathbf{R}_k, \mathbf{R}_i)\}.$$

(2.44)

If the polarizabilities are isotropic as in eqn. (2.35), this expression reduces to a form

$$V_3(ijk) = 32\pi^2\hbar \int_0^\infty d\xi\, \alpha_i(i\xi)\alpha_j(i\xi)\alpha_k(i\xi)$$
$$\times \text{Tr}\,\{\mathscr{G}(\mathbf{R}_i, \mathbf{R}_j)\mathscr{G}(\mathbf{R}_j, \mathbf{R}_k)\mathscr{G}(\mathbf{R}_k, \mathbf{R}_i)\}. \quad (2.45)$$

In the non-retarded limit $\mathscr{G}(\mathbf{R}_2, \mathbf{R}_1)$ is independent of the frequency, since terms like ξ^2/c^2 can be dropped for $c \to \infty$. Then it can be shown that

$$\lim_{c \to \infty} \text{Tr}\,\{\mathscr{G}(\mathbf{R}_i, \mathbf{R}_j)\mathscr{G}(\mathbf{R}_j, \mathbf{R}_k)\mathscr{G}(\mathbf{R}_k, \mathbf{R}_i)\}$$

$$= \frac{1}{(4\pi)^3} \frac{3}{(R_{ij}R_{jk}R_{ki})^3} [3\cos\theta_{ij}\cos\theta_{jk}\cos\theta_{ki} + 1], \quad (2.46)$$

where θ_{ij} is the angle between R_{ij} and R_{jk} in the triangle formed by R_{ij}, R_{jk}, R_{ki}. Hence

$$V_3(ijk) = \frac{3\hbar}{2\pi} \frac{1}{(R_{ij}R_{jk}R_{ki})^3} [3\cos\theta_{ij}\cos\theta_{jk}\cos\theta_{ki} + 1]$$
$$\times \int_0^\infty d\xi\, \alpha_i(i\xi)\alpha_j(i\xi)\alpha_k(i\xi), \quad (2.47)$$

which was first given by Axilrod and Teller (1943). The corresponding retarded expression is essentially eqn. (2.45) without taking this limit of $c \to \infty$, and is identical with that derived by Aub and Zienau (1960) from quantum electrodynamics.

Since very complete discussions are available in the literature (Margenau and Kestner, 1971) of three-body and many-body dispersion forces, we shall not elaborate on this aspect any further. It may be noted from eqn. (2.43) that the assumption of pairwise additivity of the dispersion interaction energy of a many-particle system simply does not hold. The corrections due to such many-body effects are important when cohesive energies are computed accurately, although two-body forces provide the dominant contribution [cf. §4.11].

The approach followed in the analysis of this section is based on consideration of the zero-point energy of the electromagnetic field in so far as it is affected by the presence of molecules. There have been a number of papers in recent literature which deal with the problem of dispersion forces from various semiclassical points of view. Mention may be made, in particular, of the work of Renne and Nijboer (1967, 1968, 1970) and Renne (1971a, b) who have considered the problem of dispersion forces in an assembly of harmonic oscillators by coupling them using the Green functions of the electromagnetic field in the non-retarded and retarded regimes, and have evaluated the interaction energy from the frequency shifts. In principle, the approach of Renne is equivalent to that followed here. We have preferred the field approach here because of its directness, and because, as will be demonstrated in Chapter 4, it lends itself readily to generalizations that take into account the finite size of the molecules in a natural way. The same problem has also been discussed in a series of papers by Boyer (1972a, 1972b, 1973) in a formalism that resembles the one followed here.

2.4 Interactions between macroscopic bodies in a continuum theory—surface modes

The dispersion interaction between two macroscopic bodies from the field point of view can be evaluated from eqn. (2.40) in a manner analogous to what has been done in §1.8. This will be discussed in some detail in Chapter 4, where the role of the size of the constituent molecules will also be considered. We shall indicate here a method of analysis of dispersion forces based on the nature of the surface electromagnetic modes excited on the macroscopic bodies, originally suggested by van Kampen, Nijboer and Schram (1968). The main idea of this approach is based on the fact that the change in the electromagnetic modes due to the presence of macroscopic homogeneous dielectric bodies can be expressed in terms of the surface modes excited on them. We shall discuss this in the non-retarded limit for simplicity, although the retarded situation can be dealt with by a straightforward generalization of the procedure.

In the non-retarded limit the electric field is obtained from the potential which satisfies Laplace's equation with the boundary conditions of continuity of the potential and of $\varepsilon(\nabla \phi)_\perp$ across the dielectric interfaces, $-(\nabla \phi)_\perp$ being the normal component of the field. Since the dielectric constants depend on the frequency, these boundary conditions can be satisfied only for specific values of the frequency which governs the time dependence of the potential. To obtain these frequencies, one can either solve for ϕ from Laplace's equation and use the boundary conditions to obtain the secular equation, or equivalently, evaluate the Green function and investigate the position of its poles in the ω-plane. For many geometrical shapes of macroscopic bodies constituted of homogeneous dielectrics with sharp boundaries the Green function can be obtained in terms of induced surface charges, and its poles would refer to the frequencies of the surface modes associated with fluctuations of these induced charges.

To illustrate this method, let us consider the case of two parallel semi-infinite dielectric slabs of dielectric constant ε_1 and ε_3, separated by a slab of dielectric constant ε_2 and thickness l (cf. Fig. 2.1). The Green function for Laplace's equation in this case has been worked out in Appendix A, and has the form

$$G(\mathbf{r}, \mathbf{r}') = -\frac{H(z')}{4\pi}\left(\frac{1}{|\mathbf{r}-\mathbf{r}'|} + \int_0^\infty \frac{d\kappa\, J_0(\kappa|\boldsymbol{\rho}-\boldsymbol{\rho}'|)F(\kappa; z, z')}{(1-\Delta_{12}\Delta_{32}e^{-2\kappa l})}\right), \quad (2.48)$$

where

$$H(z') = \left(\frac{\theta(-z')}{\varepsilon_1} + \frac{\theta(z')\theta(-z')}{\varepsilon_2} + \frac{\theta(z'-l)}{\varepsilon_3}\right) \quad (2.49a)$$

$$\boldsymbol{\kappa} = k_x\hat{e}_x + k_y\hat{e}_y \quad (2.49b)$$

$$\boldsymbol{\rho} = x\hat{e}_x + y\hat{e}_y \quad (2.49c)$$

$$\Delta_{ij} = \frac{\varepsilon_i - \varepsilon_j}{\varepsilon_i + \varepsilon_j}. \quad (2.49d)$$

Here $\theta(z)$ is the step-function, $\hat{e}_x$ and $\hat{e}_y$ are unit vectors, and $F(\kappa; z, z')$ is a complicated function whose form need not concern us here. It will be noticed that there are poles associated with the equations

$$\varepsilon_j(\omega) = 0; \quad j = 1, 2, 3. \quad (2.50)$$

These refer to bulk modes and they do not depend on the separation of the slabs. The poles arising from the roots of the equation

$$1 - \Delta_{12}(\omega)\,\Delta_{32}(\omega)e^{-2\kappa l} = 0 \quad (2.51)$$

refer to the surface modes—this is clear from the manner of derivation of the Green function in Appendix A. The roots of (2.51) give ω as a function of the (two-dimensional) wave vector $\boldsymbol{\kappa}$; i.e., the dispersion relation for the surface modes. The interaction energy per unit area is then given by summing over the zero-point energies of each ω for each value of κ, so that the energy of interaction per unit area is

$$E_{12}(l) = \sum_{\kappa} \frac{1}{2\pi i} \oint d\omega \, \frac{\hbar\omega}{2} \frac{d}{d\omega} \ln\left(1 - \Delta_{12}(\omega)\,\Delta_{32}(\omega)e^{-2\kappa l}\right)$$

$$= -\frac{\hbar}{4\pi i} \oint d\omega \int \frac{d^2\kappa}{(2\pi)^2} \ln\left(1 - \Delta_{12}(\omega)\,\Delta_{32}(\omega)e^{-2\kappa l}\right)$$

$$= \frac{\hbar}{16\pi^2 l^2} \int_0^\infty d\xi \int_0^\infty x \, dx \, \ln\left(1 - \Delta_{12}(i\xi)\,\Delta_{32}(i\xi)e^{-x}\right). \qquad (2.52)$$

In (2.52) the second line is obtained by an integration by parts and replacing $\sum_{\kappa} \equiv \int d^2\kappa/(2\pi)^2$, and the third line is obtained by integrating along the imaginary axis. (The replacement of the integral $\int_{-\infty}^{\infty} d\xi$ by $2\int_0^\infty d\xi$ is ultimately a consequence of the time reversal symmetry of Maxwell's equations.) This result (eqn. (2.52)) is the well-known expression of Lifshitz for the non-retarded interaction energy of two dielectric slabs, recovered already in §1.8 by a different method.

A more direct and simpler derivation of the secular equation is to assume that the potential $\phi(\mathbf{r}, t)$ which satisfies Laplace's equation has the form

$$\phi(\mathbf{r}, t) = f(z)e^{i(\mathbf{\kappa}\cdot\mathbf{\rho} - \omega t)}. \qquad (2.53)$$

Substituting this form in Laplace's equation we get

$$\frac{d^2 f}{dz^2} - \kappa^2 f = 0. \qquad (2.54)$$

Using the well-behavedness criterion of the solutions, we must have solutions of the form

$$\begin{aligned} f(z) &= Ae^{\kappa z}, & z &< 0 \\ &= Be^{\kappa z} + Ce^{-\kappa z}; & 0 &< z < l \\ &= De^{-\kappa z}, & z &> l. \end{aligned} \qquad (2.55)$$

The boundary conditions give us the equations

$$A = B + C$$
$$Be^{\kappa l} + Ce^{-\kappa l} = De^{-\kappa l}$$
$$\varepsilon_1 \kappa A = (\kappa B - \kappa C)\varepsilon_2$$
$$\varepsilon_2 \kappa (Be^{\kappa l} - Ce^{-\kappa l}) = -\kappa De^{-\kappa l}\varepsilon_3. \qquad (2.56)$$

The condition for solution of this set of four equations for the four unknowns A, B, C and D is the secular equation, which is

$$
\begin{vmatrix}
1 & -1 & -1 & 0 \\
0 & e^{\kappa l} & e^{-\kappa l} & -e^{-\kappa l} \\
\varepsilon_1 \kappa & -\varepsilon_2 \kappa & \varepsilon_2 \kappa & 0 \\
0 & \varepsilon_2 \kappa e^{\kappa l} & -\varepsilon_2 \kappa e^{-\kappa l} & \varepsilon_3 \kappa e^{-\kappa l}
\end{vmatrix} = 0. \tag{2.57}
$$

With some further simplification we recover again eqn. (2.51) for the surface modes.*

This kind of approach has been extended to include retardation effects and the effects of non-zero temperature (Ninham, Parsegian and Weiss, 1970; Gerlach, 1971; Nijboer and Renne, 1971; Schram, 1973; and Langbein, 1973a) in the interaction of dielectric slabs. It has also been used to work out the non-retarded interaction between dielectric spheres and cylinders and in principle can be applied to other geometries. We consider the effect of geometry in Chapter 5. The extension to include retardation requires the solution of the full Maxwell equations rather than the simpler Laplace equation which holds when $c \to \infty$. For the moment, we shall note that at finite temperature the interaction energy becomes a free energy of interaction. For two slabs, the result is

$$
F(l, T) = \frac{k_B T}{8 \pi l^2} \sum_{n=0}^{\infty}{}' I(\xi_n, l), \tag{2.58}
$$

where

$$
I(\xi_n, l) = \left(\frac{2\xi_n l \sqrt{\varepsilon_2}}{c} \right)^2 \int_1^{\infty} p \, dp \{ \ln (1 - \bar{\Delta}_{21}^R \bar{\Delta}_{23}^R \exp (-2p\xi_n l \sqrt{\varepsilon_2}/c))
$$
$$
+ \ln (1 - \Delta_{21}^R \Delta_{23}^R \exp (-2p\xi_n l \sqrt{\varepsilon_2}/c)) \} \tag{2.59}
$$

and

$$
\bar{\Delta}_{21}^R = \frac{s_1 \varepsilon_2 - p\varepsilon_1}{s_1 \varepsilon_2 + p\varepsilon_1}, \qquad \Delta_{21}^R = \frac{s_1 - p}{s_1 + p},
$$

$$
s_1 = \sqrt{p^2 - 1 + \varepsilon_1/\varepsilon_2}, \qquad \varepsilon = \varepsilon(i\xi_n). \tag{2.60}
$$

The prime on the summation symbol indicates that the term in $n = 0$ is to be taken with a factor $\frac{1}{2}$, and the sum is to be taken over imaginary frequencies $\omega_n = i\xi_n$, where $\xi_n = 2\pi n k_B T/\hbar$. This somewhat complicated

---

* In this simplification it is necessary to divide eqn. (2.57) by $(\varepsilon_1 + \varepsilon_2)$ and $(\varepsilon_3 + \varepsilon_2)$. The zeros of these two terms give the frequencies of surface modes when only (12) or (32) interfaces are present. These do not depend on l and hence do not affect the interaction energy.

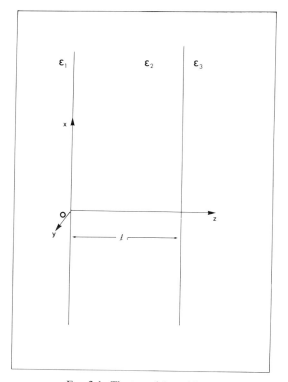

FIG. 2.1. The two-slab problem.

expression is the famous result obtained by Lifshitz* (1955) for the case that the intervening medium "2" is a vacuum, by a different method, and in the presence of an intervening medium by Dzyaloshinskii, Lifshitz and

---

* In 1894, P. N. Lebedeff (1894) wrote: "... Of special interest and difficulty is the process which takes place in a physical body when many molecules interact simultaneously, the oscillations of the latter being interdependent owing to their proximity. If the solution of this problem ever becomes possible we shall be able to calculate in advance the values of the intermolecular forces due to molecular inter-radiation, deduce the laws of their temperature dependence, solve the fundamental problem of molecular physics whether all the so-called 'molecular forces' are confined to the already known mechanical action of light radiation, to electromagnetic forces, or whether some forces of hitherto unknown origin are involved...." It seems especially fitting that his speculations and grand vision concerning the electromagnetic origin of molecular forces should have been confirmed by the Russians in the dramatic simultaneous advance, in theory by Lifshitz (1955), and in experiment by Deryaguin and Abrikosova (1956). It is even more fitting that the great Russian physicist Deryaguin should have been responsible for these advances, for Deryaguin is Lebedeff's step-son.

Pitaevskii (1961), who used the formidable apparatus of the Matsubara–Fradkin–Green function technique of quantum statistical mechanics. We shall examine the nature of the formula in detail below. If the media have also magnetic susceptibilities μ_j, $j = 1, 2, 3$, eqns. (2.59) and (2.60) have a slightly more complicated form. In that case (Richmond and Ninham, 1971), in (2.59) $\sqrt{\varepsilon_2}$ should be replaced by $\sqrt{\varepsilon_2\mu_2}$, and eqn. (2.60) becomes

$$\bar{\Delta}_{21}^{R} = \frac{s_1\varepsilon_2 - p\varepsilon_1}{s_1\varepsilon_2 + p\varepsilon_1}, \qquad \Delta_{21}^{R} = \frac{s_1\mu_2 - p\mu_1}{s_1\mu_2 + p\mu_1}. \tag{2.61}$$

The equivalence of the surface mode approach for deriving this result and the method followed by Dzyaloshinskii, Lifshitz and Pitaevskii has been discussed by Davies (1971). Apart from the original method of Lifshitz (1955), and that of Dzyaloshinskii, Lifshitz and Pitaevskii (1961), several other approaches to the general problem of electromagnetic interactions, more or less equivalent, have been developed by various authors. We have not attempted to be exhaustive in our citations. The papers of Boyer (1968a, b; 1969a, b; 1972a, b), the work of McLachlan (1963a, b, c; 1964; 1965), of Linder (1967), Buckingham (1967), Bullough (1970), and quite distinctly different approaches developed by Langbein (1973b, 1974) especially deserve note. The last reference contains a useful set of references to these different approaches.

2.5 Boundary effects on dispersion forces

A problem of some interest is the change in the dispersion interaction between two molecules when they are in close proximity to a macroscopic body. This type of effect would show up, for instance, in the lateral interaction between adsorbed molecules on a surface. One formulation of this problem has been discussed by Mahanty and Ninham (1973a). We shall briefly outline the approach here.

The starting point of this analysis can be eqn. (2.34) which we shall adapt to isotropic molecules,

$$V(R_{12}) = -8\pi\hbar \int_0^\infty d\xi\, \alpha_1(i\xi)\alpha_2(i\xi)$$
$$\times \mathrm{Tr}\,\{\mathscr{G}(\mathbf{R}_1, \mathbf{R}_2; i\xi)\mathscr{G}(\mathbf{R}_2, \mathbf{R}_1; i\xi)\}. \tag{2.62}$$

The effect of the proximity of other bodies will appear through the Green functions which will be modified from their free space form due to boundary effects. The actual evaluation of Green functions for specific geometrical shapes of the boundaries of the macroscopic bodies is computationally rather cumbersome, but in principle can be done. We shall give

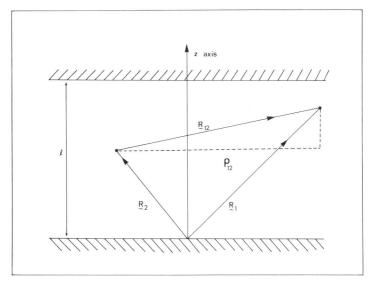

FIG. 2.2. Molecules between two plates.

here the results when the two interacting molecules are between two ideally conducting plates for illustration of the method. The separation between the plates is taken to be l (Fig. 2.2).

In the non-retarded limit when both l and R_{12} are less than the characteristic wavelength associated with electronic transitions in the molecules, $\mathscr{G}(\mathbf{r}, \mathbf{r}')$ is obtained from the electrostatic Green function

$$\mathscr{G}(\mathbf{r}, \mathbf{r}') = \nabla_r \nabla_r G(\mathbf{r}, \mathbf{r}'). \tag{2.63}$$

Since the tangential components of the electric field vanish on the plates, we can write $G(\mathbf{r}, \mathbf{r}')$ in the form

$$G(\mathbf{r}, \mathbf{r}') = -\frac{1}{2\pi^2 l} \sum_{n=1}^{\infty} \sin\left(\frac{n\pi z}{l}\right) \sin\left(\frac{n\pi z'}{l}\right) \int \frac{\mathrm{d}^2\kappa\, e^{i\boldsymbol{\kappa}\cdot(\boldsymbol{\rho}-\boldsymbol{\rho}')}}{\kappa^2 + n^2\pi^2/l^2}. \tag{2.64}$$

When the lateral distance $\rho \equiv |\boldsymbol{\rho}_1 - \boldsymbol{\rho}_2|$ between the molecules satisfies $\rho \gg l$, a suitable form of the above expression is

$$G(\mathbf{r}, \mathbf{r}') = -\frac{1}{\pi l} \sum_{n=1}^{\infty} K_0\left(\frac{n\pi\rho}{l}\right) \sin\left(\frac{n\pi z}{l}\right) \sin\left(\frac{n\pi z'}{l}\right), \tag{2.65}$$

where K_0 is a modified Bessel function in standard notation. For large ρ/l it is sufficient to retain only the term in $n = 1$ in this series, and to use the asymptotic form of K_0. Substituting this in eqn. (2.62) we get

$$V(R_{12}) \approx -\left(8\pi\hbar\int_0^\infty d\xi\, \alpha_1(i\xi)\alpha_2(i\xi)\right)$$

$$\times \frac{\pi^3}{l^6}\cos^2\left(\frac{\pi(z_1+z_2)}{l}\right)\frac{e^{-2\pi\rho/l}}{(2\pi\rho/l)}. \tag{2.66}$$

Thus the interaction energy is much weakened compared with the London ($1/R^6$) result for free space.

For $l \gg \rho$, the Green function can be cast into a different form

$$G(\mathbf{r},\mathbf{r}') = -\frac{1}{4\pi}\int_0^\infty d\kappa\, \frac{J_0(\kappa\rho)}{\sinh(\kappa l)}[\cosh\{\kappa(l-|z-z'|)\}$$

$$-\cosh\{\kappa(l-|z+z'|)\}]. \tag{2.67}$$

The analysis using this form of the Green function for different relative values of ρ and z is rather involved and details may be seen in Mahanty and Ninham (1973a). We give the results for several cases.

For $(z_1+z_2-l) \ll l$, that is when the molecules are nearly at the mid-plane of the two conducting plates,

$$V(R_{12}) \approx V_{\text{London}}(R_{12})\left\{1+\tfrac{5}{6}\zeta(3)\left(\frac{R_{12}}{l}\right)^3\left(\frac{3(z_1-z_2)^2}{R_{12}^2}-1\right)+\cdots\right\}, \tag{2.68}$$

where $V_{\text{London}}(R_{12})$ is the London interaction in free space and $\zeta(3)$ a Riemann zeta function. It may be noted on the mid-plane, i.e., when $z_1 = z_2$ the interaction is diminished from the London result. But when $|z_1-z_2| > (\rho_{12}/\sqrt{2})$ the interaction is enhanced from the free space value. For $(z_1+z_2) \ll l$, that is the molecules are close to one of the conducting planes, we have

$$V(R_{12}) \approx V_{\text{London}}(R_{12})\{\tfrac{2}{3}(1-\zeta(3)\rho_{12}^3/l^3)\} \tag{2.69}$$

which implies a reduction from the London value by a factor of $\tfrac{2}{3}$.

In the retarded region the form of the Green function is rather complicated, and again we refer to the original paper for details. One interesting result which emerges for $\rho_{12} \gg l$ is

$$V(R_{12}) \approx -\frac{192}{\pi}\hbar c\, \frac{\alpha_1(0)\alpha_2(0)}{l^2\rho_{12}^5}. \tag{2.70}$$

This indicates a marked increase in the interaction energy when compared with the Casimir–Polder ($1/R^7$) form when the molecules are between parallel conducting plates.

The approach outlined here can easily be generalized to take account of other types of boundaries, such as dielectric boundaries. We shall not dwell on this point further at this stage, except to make a general observation that when two molecules are near a dielectric surface, the non-retarded dispersion interactions between them will be diminished from its free space value. (Physically one would expect this because the molecules interact not only with each other, but also with their images. The net effect is to make the interaction look more like an induced multipole–multipole interaction rather than an induced dipole–dipole interaction.) On the other hand, in the retarded region there may be an increase from the free space value if guided electromagnetic waves can propagate along the surface.

2.6 A note on interactions between excited molecules

The interaction between two identical molecules, one of which is in the ground state and the other in an excited state, the so-called resonance interaction, is not a dispersion interaction in the sense of arising out of the change in the zero-point energy of the field. It arises due to the radiation field generated by the excited molecule undergoing a transition to the ground state, this radiation field affecting the polarization of the other molecule. The problem of evaluation of this interaction has been comprehensively reviewed by Margenau and Kestner (1971). We shall indicate here briefly how one can formulate the problem of this resonance interaction between two molecules semi-classically, in a manner closely analogous to the semi-classical approach developed in this chapter.

The interaction energy here is basically the energy transferred from the excited molecule to the one in its ground state. Since the radiation due to electronic transitions can be represented equivalently as arising out of oscillating dipoles, the transferred energy can be interpreted as the energy of one oscillating dipole in the radiation field of the other (McLone and Power, 1964). If we consider an oscillating dipole source at $\mathbf{R}_1$ with frequency ω_0, the electric field due to radiation from the dipole can be written as

$$\mathbf{E}(\mathbf{r}, t) = -\frac{4\pi\omega_0^2}{c^2}\mathcal{G}(\mathbf{r}, \mathbf{R}_1; \omega_0)\mathbf{P}_1 e^{i\omega_0 t}, \qquad (2.71)$$

where $\mathbf{P}_1$ is the dipole moment amplitude of the dipole. This follows from an analysis similar to that which leads to eqn. (2.28).

The energy in the resulting electric field of another oscillating dipole $\mathbf{P}_2$

of the same amplitude and frequency $(-\omega_0)$ located at $\mathbf{R}_2$ is

$$E(R_{12}) = \{\mathbf{P}_2\mathbf{E}(\mathbf{R}_2, t)\}e^{-i\omega_0 t}$$

$$= \frac{4\pi\omega_0^2}{c^2}[\mathbf{P}_2\mathcal{G}(\mathbf{R}_2, \mathbf{R}_1; \omega_0)\mathbf{P}_1]. \tag{2.72}$$

The change of sign in ω_0 for the second molecule is consistent with the quantum mechanical picture that it absorbs the quantum whereas the first one emits it. The dipole moment amplitudes can be worked out quantum mechanically from the electronic wave functions of the molecules.

In free space $\mathcal{G}(\mathbf{R}_2, \mathbf{R}_2; \omega_0)$, as obtained from eqns. (2.29) and (2.30) has the limiting forms,

$$\mathcal{G}(\mathbf{R}_2, \mathbf{R}_1; \omega_0) \propto 1/R_{12}^3 \quad \text{in the non-retarded region;}$$

$$\propto \frac{e^{i\omega_0 R_{12}/c}}{R_{12}} \quad \text{in the retarded region.}$$

We thus find the results that the interaction between two identical molecules when one of them is excited varies as $1/R^3$ in the non-retarded region and $[\cos(\omega_0 R/c)/R]$ in the retarded region.

When the molecules are in a region where there are macroscopic bodies present, the structure of $\mathcal{G}(\mathbf{R}_2, \mathbf{R}_1; \omega_0)$ will be affected by boundary effects. These may lead to substantial changes in the R-dependence of the resonance interaction. If the molecules are in wave guides, for instance, in the retarded region there is a considerable enhancement in the interaction (Mahanty and Ninham, 1973b). This sort of enhancement will have implications in resonance energy transfer in inhomogeneous systems.

2.7   Temperature dependence of dispersion interactions

An important aspect of dispersion interaction between molecules or macroscopic bodies to which we have briefly alluded in §2.4 is the influence of temperature on it. As we have seen, the interaction arises due to changes in the modal frequencies of the electromagnetic field due to the presence of the bodies. At a finite temperature the energy associated with each frequency ω is not just the zero-point energy $\hbar\omega/2$, but rather the internal energy

$$\left\{\frac{\hbar\omega}{2}\coth\left(\frac{\hbar\omega}{2k_BT}\right)\right\}.$$

More often in practice one needs the Helmholtz free energy, which for an oscillator of frequency ω_j is $k_BT\ln[2\sinh(\hbar\omega_j/2k_BT)]$. If we sum such expressions over the change in modal frequencies, we will obviously have a temperature dependence of the corresponding energies and free energies, which has been included in eqn. (2.58). This temperature dependence is additional to the implicit temperature dependence of the dielectric susceptibilities which occur in the general formula. We shall examine the nature of temperature-dependent interaction energies in detail in Chapter 3. At this point we shall merely indicate the steps which lead to the form eqn. (2.58) for the Helmholtz free energy.

If we obtain a secular equation

$$D(\omega) = 0 \tag{2.73}$$

for the change in the modal frequencies [such as eqn. (2.31) or eqn. (2.51)], the dispersion free energy can be defined as

$$F(l, T) = \frac{1}{2\pi i} \oint d\omega \, k_B T \ln \left\{ 2 \sinh \left(\frac{\hbar \omega}{2 k_B T} \right) \right\} \frac{d}{d\omega} \ln D(\omega). \tag{2.74}$$

The contour excludes singularities other than those of $D(\omega)$. Since

$$q(\omega) = k_B T \ln \left(2 \sinh \frac{\hbar \omega}{2 k_B T} \right)$$

has branch points it is convenient to expand it as

$$q(\omega) = \frac{\hbar \omega}{2} - k_B T \sum_{n=1}^{\infty} \left\{ \exp \left(-\frac{n \hbar \omega}{k_B T} \right) \right\} n \tag{2.75}$$

and consider each term separately. Proceeding formally, we choose the contour again as in Fig. 1.7 to be a path along the imaginary axis from $i\infty$ to $-i\infty$ and around the right half ω-plane along a semi-circular path with infinite radius. Since the contribution from the semicircle can be shown to vanish, we can write

$$F(l, T) = \frac{1}{2\pi i} \int_{\infty}^{-\infty} q(i\xi) \frac{d}{d\xi} (\ln D(i\omega)) \, d\xi$$

$$= \frac{\hbar}{4\pi} \int_{-\infty}^{\infty} \ln D(i\xi) \, d\xi + \frac{\hbar}{2\pi} \sum_{n=1}^{\infty} \int_{-\infty}^{\infty} \cos \left(\frac{n \hbar \xi}{k_B T} \right) \ln D(i\xi) \, d\xi$$

$$- \frac{\hbar i}{2\pi} \sum_{n=1}^{\infty} \int_{-\infty}^{\infty} \sin \left(\frac{n \hbar \xi}{k_B T} \right) \ln D(i\xi) \, d\xi. \tag{2.76}$$

The last line follows by substituting (2.75) into the first line, and integrating by parts. The last term of (2.76) is pure imaginary and vanishes. The summation over cosine terms can be carried out by interchanging summation and integration and making use of the identity

$$\sum_{n=1}^{\infty} \cos nx = \pi \sum_{n=-\infty}^{\infty} \delta(x - 2\pi n) - \tfrac{1}{2}. \qquad (2.77)$$

When the delta functions are substituted into the integrals, the integrations can be carried out trivially leading to the result

$$F(l, T) = \frac{k_B T}{2} \sum_{n=-\infty}^{\infty} \ln D(i\xi_n); \qquad \xi_n = \frac{2\pi n k_B T}{\hbar}. \qquad (2.78)$$

In the non-retarded limit $D(i\xi_n)$ is an algebraic function of ε_1, ε_2 and ε_3, so that* if

$$\varepsilon(\omega) = 1 + \sum_j \frac{c_j}{\omega_j^2 - \omega^2},$$

then $D(i\xi_n) = D(-i\xi_n)$, and

$$F(l, T) = k_B T \sum_{n=0}^{\infty}{}' \ln D(i\xi_n). \qquad (2.79)$$

2.8 Pairwise summation as a special case

We shall discuss the nature of retardation for the two slab problem in the following section. For the moment it suffices to state that the complicated equations (2.58)–(2.60) reduce to eqn. (2.52) in the non-retarded limit. We can then make an easy connection with formulae obtained by pairwise summation. Notice already that we have made two assumptions, that the temperature is zero, and the velocity of light infinite. If we expand the logarithm of eqn. (2.52) and integrate term by term we have

$$E(l) = -\frac{\hbar}{16\pi^2 l^2} \int_0^{\infty} d\xi \sum_{n=1}^{\infty} \frac{(\Delta_{12}\Delta_{32})^n}{n^2}. \qquad (2.80)$$

Terms higher than the first can be ignored for all practical purposes if we deal with dispersion forces alone. We have then the approximation

$$E(l) \approx -\frac{\hbar}{16\pi^2 l^2} \int_0^{\infty} d\xi \, \Delta_{12}\Delta_{32}; \qquad \Delta_{ij} = \frac{\varepsilon_i(i\xi) - \varepsilon_j(i\xi)}{\varepsilon_i(i\xi) + \varepsilon_j(i\xi)} \qquad (2.81)$$

* In the retarded case $\ln D(i\xi)$ is a complicated function which has branch points. Translation of the contour in the manner indicated is not formally allowed. The argument can be made rigorous by a convenient artifice [see Schram, 1973].

To exhibit the connection of this approximation with pairwise summation, we consider several limiting cases.

Suppose medium 2 is a vacuum and media 1 and 3 are identical non-polar gases. Then by the Lorentz formula $\varepsilon = 1 + 4\pi N\alpha$, where N is the density of molecules. If the polarizability of a molecule is given by the form

$$\alpha_1 = \frac{\alpha_1(0)}{1 - \omega^2/\omega_1^2} \tag{2.82}$$

we have from (2.81)

$$E(l) \approx -\frac{\hbar}{16\pi^2 l^2} \int_0^\infty \left(\frac{16\pi^2 N^2 \alpha_0^2}{4}\right) \frac{d\xi}{(1 + \xi^2/\omega_1^2)^2}$$

$$= -\frac{A_{11}}{12\pi l^2}; \qquad A_{11} \equiv \pi^2 N^2 \left(\frac{3\hbar\omega_0}{4}\alpha_1^2(0)\right) \tag{2.83}$$

which is in agreement with the Hamaker result. To obtain this expression we have put the denominator in Δ_{12}, $\varepsilon_1 + \varepsilon_2 \approx 2$. Two-body forces are proportional to $\alpha^2(0)$, and three-body forces to $\alpha^3(0)$. Consequently we see that dropping the dependence on the denominators of eqn. (2.81) is tantamount to dropping 3-body and, in general, n-body interactions.

If the three media are different, and have polarizabilities given by (2.82) repetition of the process gives

$$E(l) = -\frac{1}{12\pi l^2}\{A_{13} + A_{22} - A_{23} - A_{12}\}, \tag{2.84}$$

where

$$A_{ij} = \pi^2 N_i N_j \frac{3\hbar}{2} \alpha_i(0)\alpha_j(0) \frac{\omega_i \omega_j}{\omega_i + \omega_j}. \tag{2.85}$$

If the interacting media are identical $(1 \equiv 3)$ (2.84) reduces to

$$E(l) = -\frac{1}{12\pi l^2}(A_{11} + A_{22} - 2A_{12}), \tag{2.86}$$

again in agreement with Hamaker. Throughout the literature on colloid science there has been some confusion about this result, and the usual prescription for calculation of "Hamaker constants" has been to divide eqn. (2.86) by a factor $\varepsilon_2 = n^2$, where n is the refractive index of medium 2 in order to allow for the transmission of the force through a dielectric. If we compare with the proper expression eqn. (2.81), we see that this prescription and (2.86) are both wrong. This is an apparently trivial point, but deserves comment. For many practical problems where a knowledge of interaction energy enters, specific interaction properties are determined primarily by electrostatic double-layer forces, and depend sensitively on ionic strength, pH and other factors. For such systems, a gross description of attractive energies is sufficient. For these situations, one can say that Hamaker theory "works." On the other hand, one interesting class of problems to be discussed subsequently depend on a most delicate balance of two- and three-body effects, and are very sensitive to the dielectric properties of the media. The use of approximate formulae like (2.86) can be misleading.

2.9 The nature of retardation in Lifshitz theory—explicit analysis

Returning to the general formula of eqn. (2.58), we now ask what happens when the distance between two slabs becomes large. First consider the case of zero temperature. Following Lifshitz (1955), the sum can be replaced by an integral through the transformation

$$k_B T \sum_{n=0}^{\infty}{}' f(\xi_n) \underset{T \to 0}{\Rightarrow} \frac{\hbar}{2\pi} \int_0^\infty d\xi f(\xi), \tag{2.87}$$

where the higher order terms at finite T can be found using Poisson's or the Euler–Maclaurin summation formula. We then have

$$F(l, 0) \equiv E(l) = \frac{\hbar}{16\pi^2 l^2} \int_0^\infty I(\xi, l) \, d\xi. \tag{2.88}$$

The change of variable $x = 2pl\xi/c$ gives the alternative form (we consider identical media $1 \equiv 3$)

$$E(l) = \frac{\hbar c}{32\pi^2 l^3} \int_0^\infty dx \int_1^\infty dp \frac{x^2}{p^2} \varepsilon_2 \{\ln(1 - (\bar{\Delta}_{12}^R)^2 \exp(-x\sqrt{\varepsilon_2}))$$
$$+ \ln(1 - (\Delta_{12}^R)^2 \exp(-x\sqrt{\varepsilon_2}))\}, \tag{2.89}$$

where now $\varepsilon = \varepsilon(icx/2pl)$. At very large distances because of the exponential factors in the integrand in x, the only contributing region of importance will be where $x \approx 1/\sqrt{\varepsilon_2} \lesssim 1$, and since $p \geq 1$, the argument of the functions $\varepsilon(icx/2pl)$ is close to zero throughout this region. Consequently we can replace ε_1, ε_2 by their values at $\xi = 0$, i.e., by the electrostatic dielectric constants. After making this approximation and the further change of variable $x \to x/\sqrt{\varepsilon_2(0)}$ we have

$$E(l) \approx \frac{\hbar c}{32\pi^2 l^3} \int_0^\infty dx \int_1^\infty dp \frac{x^2}{p^2} \{\ln(1 - (\bar{\Delta}_{12}^R(0))^2 \exp(-x))$$
$$+ \ln(1 - (\Delta_{12}^R(0))^2 \exp(-x))\}, \tag{2.90}$$

where the Δ's are given by eqn. (2.60) and

$$s(0) = \sqrt{\frac{\varepsilon_1(0)}{\varepsilon_2(0)} - 1 + p^2}.$$

For two metallic plates separated by a vacuum, we can put $\varepsilon_2 = 1$, $\varepsilon_1(0) = \infty$, $\Delta_{12}(0) = \bar{\Delta}_{12}(0) = 1$, and

$$
\begin{aligned}
E(l) &\approx \frac{\hbar c}{16\pi^2 l^3} \int_0^\infty x^2 \, dx \int_1^\infty \frac{dp}{p^2} \ln\left[1 - \exp(-x)\right] \\
&= -\frac{\hbar c}{16\pi^2 l^3} 2! \sum_{n=1}^\infty \frac{1}{n^4} = -\frac{\hbar c \pi^2}{720 l^3}.
\end{aligned}
\tag{2.91}
$$

This is identical with the result of Casimir for two metallic plates (eqn. (2.6)). For two identical dielectrics or a dielectric and metal separated by a vacuum the corresponding results have been given by Lifshitz numerically.

From eqn. (2.90) we can also deduce the force between two atoms at large distances. For suppose that the two media are very dilute. Then the difference $\varepsilon_1(0) - 1$ is small. After expansion of the integrand of (2.90) in powers of this small parameter, and keeping only leading terms, we have

$$
\begin{aligned}
f(l, 0) &= -\frac{\hbar c}{32\pi^2 l^3} [\varepsilon_1(0) - 1]^2 \int_0^\infty x^2 \, dx e^{-x} \int_1^\infty \frac{1 - 2p^2 + 2p^4}{8p^6} \, dp \\
&= -\frac{23\hbar c}{(1920)\pi^2 l^3} [\varepsilon_1(0) - 1]^2.
\end{aligned}
\tag{2.92}
$$

This corresponds with the result which would have been obtained by pairwise summation of the retarded individual atomic potential

$$E(R) = -\frac{23\hbar c}{4\pi R^7} \alpha_1^2(0),
\tag{2.93}$$

where $\alpha_1(0)$ is the static polarizability of the atoms, and emerges here solely from macroscopic considerations.

Temperature dependence

By replacing the sum in (2.58) by an integral, we make the approximation that the explicit temperature dependence of the interaction free energy is negligible. [There is an implicit temperature dependence since the several $\varepsilon = \varepsilon(i\xi_n)$ depend on T.] This is not always a good approximation, particularly for many biological situations. The term in $n = 0$ in the sum (2.58) behaves as $1/l^2$ for all l whereas the rest of the term, as given by (2.90), behave like $1/l^3$ for large l. Hence at some distance l, the temperature-dependent term will dominate. If we put $n = 0$ in the first term of (2.58) this term is indeterminant (because in (2.59) the factor ξ_n^2 vanishes while the integral over p is divergent). The indeterminacy can be removed by replacing p by the new variable of integration $x = 2p\xi_n l\sqrt{\varepsilon_2(0)}/c$, as a result of which the factor ξ_n^2 goes out. On putting $\xi_n = 0$, we get

$$E(l)_{n=0} = \frac{k_B T}{16\pi l^2} \int_0^\infty x \, dx \, \ln\left\{1 - \left(\frac{\varepsilon_1(0) - 1}{\varepsilon_1(0) + 1}\right)^2 e^{-x}\right\}. \tag{2.94}$$

Thus at large distances the decrease of the interaction free energy slows down and goes again like a $1/l^2$ law.

A more refined analysis

The last fact, that the interaction reverts back to a $1/l^2$ form for large l seems peculiar, and warrants further examination. The apparent anomaly arises because the approximations used to derive (2.90) give us the leading term (proportional to $1/l^3$) of an asymptotic expansion. Asymptotic expansions are not unique, and a better and more revealing representation of our function $I(\xi, l)$ can be obtained as follows (Ninham and Parsegian, 1970): Consider eqn. (2.59). To a good approximation the logarithms which occur can be replaced by their leading terms and we consider first

$$I_{\text{approx}} = -\left(\frac{\xi}{\xi_s}\right)^2 \int_1^\infty p \, dp \left(\frac{s\varepsilon_2 - p\varepsilon_1}{s\varepsilon_2 + p\varepsilon_1}\right)^2 e^{-(\xi/\xi_s)p}, \tag{2.95}$$

where

$$\xi_s = \left(\frac{c}{2l\sqrt{\varepsilon_2}}\right).$$

For a fixed frequency ξ, the function $(s\varepsilon_2 - p\varepsilon_2)/(s\varepsilon_2 + p\varepsilon_2)$ is slowly varying as a function of p. Since the exponential is a decreasing function of p while p itself is increasing, the integrand of eqn. (2.95) has a maximum at some $p \equiv p_0$ given by

$$\frac{d}{dp}\left(\ln p - \xi\frac{p}{\xi_s}\right) = 0; \qquad p_0 \approx \frac{\xi_s}{\xi} = \frac{c}{2\xi l\sqrt{\varepsilon_2}}. \tag{2.96}$$

This means that for a fixed value of l, and small ξ, the major contribution comes from the large values of p. On the other hand for the same given value of l, for large ξ, we have $p_0 \lesssim 1$ and the main contribution to the p integral comes from the region $p \approx 1$. Two cases then arise:

(a)
$$\xi_s/\xi = \frac{c}{2\xi l\sqrt{\varepsilon_2}} \gg 1. \tag{2.97}$$

This corresponds to the non-retarded case already discussed. For these values of ξ we may put $p \approx s$ in the integrand of (2.88). [The term in Δ_{12}^R disappears here.] In the integrand which remains the lower limit can be replaced by zero and we have

$$
I(\xi, l)(\xi < \xi_s) \approx (\xi/\xi_s)^2 \int_0^\infty p \, dp \, \ln \left\{ 1 - \left(\frac{\varepsilon_2 - \varepsilon_1}{\varepsilon_2 + \varepsilon_1} \right)^2 e^{-p\xi/\xi_s} \right\}
$$

$$
= \int_0^\infty x \, dx \, \ln \left\{ 1 - \left(\frac{\varepsilon_2 - \varepsilon_1}{\varepsilon_2 + \varepsilon_1} \right)^2 e^{-x} \right\}
$$

$$
\approx \left(\frac{\varepsilon_2 - \varepsilon_1}{\varepsilon_2 + \varepsilon_1} \right)^2 . \tag{2.98}
$$

The correction terms are of the order of $(\xi/\xi_s)^2$. At a distance $l = 50 \text{ Å}$, $\xi_s \approx 3 \times 10^{16}$ rad/sec which lies in the mid ultraviolet. For $\xi > 3 \times 10^{16}$ the function $[(\varepsilon_2 - \varepsilon_1)/(\varepsilon_2 + \varepsilon_1)]^2$ is already very small compared with its maximum value (since for frequencies greater than 3×10^{16} little further absorption occurs). Hence effects of retardation will be small. On the other hand for $l = 500 \text{ Å}$, ξ_s lies in the near ultraviolet and we expect these high frequency contributions to be diminished.

(b) $$\xi_s/\xi = \frac{c}{2\xi l \sqrt{\varepsilon_2}} \ll 1.$$

Again consider eqn. (2.95). For these values of ξ the main contribution comes from values of p near 1. Writing $x = p - 1$, we have

$$
I(\xi, l)(\xi > \xi_s) \approx (\xi/\xi_s)^2 \int_0^\infty \left(\frac{s_1 - (1+x)\varepsilon_1/\varepsilon_2}{s_1 + (1+x)\varepsilon_1/\varepsilon_2} \right)^2
$$
$$
\times \exp\left[-\xi/\xi_s (1+x) \right] dx. \tag{2.99}
$$

After a further change of variable to $x\xi/\xi_s = y$, noting that the major contribution to the integral comes from $y \approx 0$, we can expand the integrand in ascending powers of y to get

$$
I(\xi, l)(\xi > \xi_s) \sim \left(\frac{\xi}{\xi_s} \right) \exp \left(-\frac{\xi}{\xi_s} \right) \int_0^\infty dy
$$

$$
\times \exp(-y) \left(\frac{\sqrt{\varepsilon_2} - \sqrt{\varepsilon_1}}{\sqrt{\varepsilon_2} + \sqrt{\varepsilon_1}} \right)^2 \left\{ 1 + 0 \left(y \frac{\xi_s}{\xi} \right) \right\}
$$

$$
= \frac{\xi}{\xi_s} \exp \left(-\frac{\xi}{\xi_s} \right) \left(\frac{\sqrt{\varepsilon_2} - \sqrt{\varepsilon_1}}{\sqrt{\varepsilon_2} + \sqrt{\varepsilon_1}} \right)^2 \left\{ 1 + 0 \left(\frac{\xi_s}{\xi} \right) \right\}. \tag{2.100}
$$

A similar contribution comes from the remaining integrand of (2.90).

Thus for $\xi > \xi_s$, $I(\xi, l)$ has a form essentially equivalent to (2.98) but with exponential damping. Complete asymptotic expansions which converge rapidly for computation can easily be constructed, but they are messy, and there is little point in so doing. It is simpler and safer to calculate numerically for each case.

We can now develop a better understanding of the way in which retardation affects interactions. Write (2.58) as

$$F(l, T) = \frac{k_B T}{8 \pi l^2} \sum_{n=0}^{\infty}{}' I(\xi_n, 0) \eta(\xi_n, l), \qquad (2.101)$$

where

$$\eta(\xi_n, l) = \frac{I(\xi_n, l)}{I(\xi_n, 0)}. \qquad (2.102)$$

Eqns. (2.98) and (2.100) tell us that the function $\eta(\xi_n, l)$ is essentially a cut-off function, multiplying the non-retarded $I(\xi_n, 0)$. Its form is exhibited for a particular model calculation in Fig. 2.3. Arrows indicate frequencies at which $\xi = c/2l$. Note that retardation damping is effectively a shift of the curve $\eta(\xi, l)$ to the left to cut out higher frequencies. The range of damping for this system is approximately one decade wide and centred around $\xi = c/2l$. As l increases, the interaction retains the $1/l^2$ dependence, but the coefficient which depends on l changes as successively higher frequencies are damped.

Evidently the effects of retardation are extremely subtle. For interactions of unlike substances, and even for like substances of different

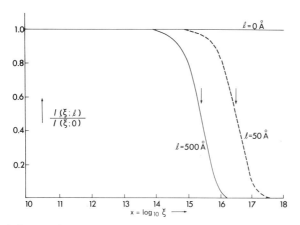

FIG. 2.3. The influence of retardation on infrared and ultraviolet contributions to van der Waals energies. Retardation factor for damping contributions at different frequencies. [From Ninham and Parsegian (1970).]

geometries, the non-retarded function $I(\xi_n, 0)$ can be either positive or negative depending on ξ_n. Hence we can expect to find situations, e.g., where the "attractive" force is at first attractive, as higher frequencies are damped out with decreasing l goes repulsive, and at smaller distances still attractive again. This is a remarkable and most delicate phenomenon, which would be quite inaccessible by the older methods of treating interactions. In the older colloid literature various approximate formulae can be found which attempt to take account of retardation. Such formulae which "patch up" pairwise summation, assuming a single adsorption frequency, should not be used for numerical work.

References

Aub, M. R. and Zienau, S. (1960). *Proc. Roy. Soc.* A**257**, 464.

Axilrod, B. M. and Teller, E. (1943). *J. Chem. Phys.* **11**, 299.

Blokhintsev, D. I. (1964). "Quantum Mechanics," D. Reidel Publ. Co., Dordrecht-Holland. §92.

Boyer, T. D. (1968a). *Phys. Rev.* **174**, 1631.

Boyer, T. D. (1968b). *Phys. Rev.* **174**, 1764.

Boyer, T. D. (1969a). *Phys. Rev.* **182**, 1374.

Boyer, T. D. (1969b). *Phys. Rev.* **186**, 1304.

Boyer, T. D. (1972a). *Phys. Rev.* A**5**, 1799.

Boyer, T. D. (1972b). *Phys. Rev.* A**6**, 314.

Boyer, T. D. (1973). *Phys. Rev.* A**7**, 1832.

Buckingham, A. D. (1967). *Adv. Chem. Phys.* **12**, 107.

Bullough, R. K. (1970). *J. Phys.* A**3**, 751, and earlier papers of the same author cited therein.

Casimir, H. B. G. (1948). *Proc. Kon. Ned. Akad. Wetensh.* **51**, 793.

Casimir, H. B. G. (1949). *J. Chim. Phys.* **46**, 407.

Casimir, H. B. G. and Polder, D. (1946). *Nature* **158**, 787.

Casimir, H. B. G. and Polder, D. (1948). *Phys. Rev.* **73**, 360.

Davies, B. (1971). *Phys. Letts.* **37**A, 391.

Davies, B. (1972). *Chem. Phys. Letts.* **16**, 388.

Deryaguin, B. V. and Abrikosova, I. I. (1956). *J. Exp. Theor. Phys.* **30**, 993; **31**, 3 [*Sov. Phys. JETP* (1957) **3**, 819, **4**, 2].

Dzyaloshinskii, I. E., Lifshitz, E. M. and Pitaevskii, L. P. (1961). *Adv. Phys.* **10**, 165.

Feinberg, G. and Sucher, J. (1970). *Phys. Rev.* A**2**, 2395.

Gerlach, E. (1971). *Phys. Rev.* B**4**, 393; see also (1971) *J. Vac. Sci. Techn.* **9**, 747.

Langbein, D. (1973a). *J. Chem. Phys.* **58**, 4476; *Solid State Comm.* **12**, 853.

Langbein, D. (1973b). *Advan. Solid State Phys. (Festkörperprobleme)* **13**, 85.

Langbein, D. (1974). "Theory of van der Waals Attraction," Springer Tracts in Modern Physics, Springer-Verlag, Berlin, **72**, 139pp.

Lebedeff, P. N. (1894). *Wied. Ann.* **52**, 621; quoted in Deryaguin, B. V., Abrikosova, I. I. and Lifshitz, E. M. (1956). *Quart. Revs. Chem. Soc.* **10**, 295.

Lifshitz, E. M. (1955). *J. Exp. Theoret. Phys. USSR* **29**, 94.

Lifshitz, E. M. (1956). *Sov. Phys. JETP* **2**, 73.

Linder, B. (1967). "Intermolecular Forces," v. 12 (ed. J. O. Hirschfelder). Adv. Chem. Phys., Interscience John Wiley.

Mahanty, J. and Ninham, B. W. (1973a). *J. Phys.* A**6**, 1140.

Mahanty, J. and Ninham. B. W. (1973b). *Phys. Letts.* **43**A, 495.

Margenau, H. and Kestner, N. R. (1971). "Theory of Intermolecular Forces," Pergamon Press, Oxford, 2nd ed.

Mavroyannis, C. and Stephen, M. J. (1962). *Mol. Phys.* **5**, 629.

McLachlan, A. D. (1963a). *Proc. Roy. Soc.* A**271**, 387.

McLachlan, A. D. (1963b). *Proc. Roy. Soc.* A**274**, 80.

McLachlan, A. D. (1963c). *Mol. Phys.* **6**, 423.

McLachlan, A. D. (1964). *Mol. Phys.* **7**, 381.

McLachlan, A. D. (1965). *Discuss. Farad. Soc.* **40**, 239.

McLone, R. R. and Power, E. A. (1964). *Mathematika* **11**, 91.

Mitchell, D. J. Ninham, B. W. and Richmond, P. (1971). *Am. J. Phys.* **40**, 674.

Nijboer, B. R. A. and Renne, M. J. (1971). *Physica Norvegica* **5**, 243.

Ninham, B. W. and Parsegian, V. A. (1970). *Biophys. J.* **10**, 646.

Ninham, B. W., Parsegian, V. A. and Weiss, G. (1970). *J. Stat. Phys.* **2**, 323.

Renne, M. J. (1971a). *Physica* **53**, 193.

Renne, M. J. (1971b). *Physica* **56**, 125.

Renne, M. J. and Nijboer, B. R. A. (1967). *Chem. Phys. Letts.* **1**, 317.

Renne, M. J. and Nijboer, B. R. A. (1968). *Chem. Phys. Letts.* **2**, 35.

Renne, M. J. and Nijboer, B. R. A. (1970). *Chem. Phys. Letts.* **6**, 601.

Richmond, P. and Ninham, B. W. (1971). *J. Phys. C.* **4**, 1988.

Schram, K. (1973). *Phys. Letts.* **43**A, 282.

Scully, M. O. and Sargent, M. (1972). *Phys. Today* **25**, 28.

van Kampen, N. G., Nijboer, B. R. A. and Schram, K. (1968). *Phys. Letts.* **26**A, 307.

Chapter 3

Calculations and Comparison of Theory with Experiment

3.1 Frequency dependence of the dielectric susceptibility

An apparent obstacle to the application of the formulae of the previous chapter is our limited knowledge of the functions $\varepsilon(\omega)$. In principle these functions contain complete information on the strength and location of the energy absorption spectrum for all frequencies from zero through to X-ray frequencies, which can be obtained theoretically only by solving the whole many-body problem. Such complete information is of course not available. Very fortunately, however, only partial knowledge of the functions $\varepsilon(\omega)$ is often sufficient to work out the force between bodies (Ninham and Parsegian, 1970b). Partial data together with known constraints on the function sharply restrict the ambiguity in using $\varepsilon(\omega)$ to calculate molecular forces.

This can be seen as follows. First of all we note that we are interested in $\varepsilon(\omega)$ as a function of the complex variable ω only on the imaginary axis $\omega = i\xi$. On the imaginary axis it can be shown that $\varepsilon(i\xi)$ is always real and monotonic decreasing (Landau and Lifshitz, 1960). Secondly, for real frequencies up into the ultraviolet $\varepsilon(\omega)$ can be written with very good approximation in the form

$$\varepsilon(\omega) = 1 + \sum_r \frac{C_r}{(1 - i\omega/\omega_r)} + \sum_j \frac{f_j}{[1 - (\omega/\omega_j)^2 - i\gamma_j\omega/\omega_j^2]}. \tag{3.1}$$

The first term describes the effect of possible Debye rotational relaxation frequencies, and the second models absorption using a Lorentz harmonic-oscillator model for the dielectric. The constants $C_r, f_j, \omega_r, \omega_j$ can be obtained from tables of dielectric data to provide a representation of $\varepsilon(\omega)$ on the real axis through the visible to the near-ultraviolet frequency range in a manner illustrated below. What are not known generally and are rather difficult to determine are the damping coefficients γ_j associated with Lorentz oscillators. This makes the estimation of $\varepsilon(\omega)$ on the real axis difficult. But on the imaginary axis, putting $\omega = i\xi$ in (3.1) we have

$$\varepsilon(i\xi) = 1 + \sum_r \frac{C_r}{(1 + \xi/\omega_r)} + \sum_j \frac{f_j}{[1 + (\xi/\omega_j)^2 + \gamma_j \xi/\omega_j^2]}. \tag{3.2}$$

Since for dielectrics the widths of the absorption spectra as measured by γ_j are always small compared with the absorption frequency ω_j, on the imaginary axis the γ_j's can be dropped to a good approximation. Beyond the ultraviolet region we have little experimental knowledge of $\varepsilon(\omega)$. However in the far-ultraviolet and soft X-ray region, all matter responds like a free electron gas, and at sufficiently high frequencies the response function $\varepsilon(\omega)$ must go over to that of an electron gas (Landau and Lifshitz, 1960):

$$\varepsilon(\omega) = 1 - \frac{\omega_p^2}{\omega^2}; \qquad \omega_p^2 = 4\pi Ne^2/m. \tag{3.3}$$

Here N, e, m are electronic number density, charge and mass, and ω_p is the plasma frequency.

On the imaginary axis the forms (3.2) and (3.3) are individually monotonic decreasing in ξ, but the regions in which each provide a valid representation of $\varepsilon(i\xi)$ do not overlap (cf. Fig. 3.1). It is sometimes convenient to construct an interpolation formula for $\varepsilon(i\xi)$ in this intermediate region. This form must satisfy the constraint that $\varepsilon(i\xi)$ be monotonic decreasing. It is then necessary to verify that the calculated dispersion energies are insensitive to the particular choice of the interpolation scheme. The uncertainty so introduced is unavoidable, however disguised, and does lead to difficulties in investigating interactions across a vacuum. Two happy circumstances conspire to reduce errors in prediction. The first is the influence of retardation. From §2.9 we see that at distances $\approx 50 \text{ Å}$, the effective cut-off frequency $\xi_s = c/2l\sqrt{\varepsilon_2}$ satisfies $\xi_s \lesssim 4 \times 10^{16}$ rad/sec and frequencies beyond the plasma frequency where the curves for $\varepsilon(i\xi)$ are uncertain (Fig. 3.1) do not contribute much to the integral for the force or energy.

Secondly, for interactions involving condensed media, the plasma frequencies of the interacting media are more nearly equal than they are in the case that one of the media is a vacuum. (Remember that the

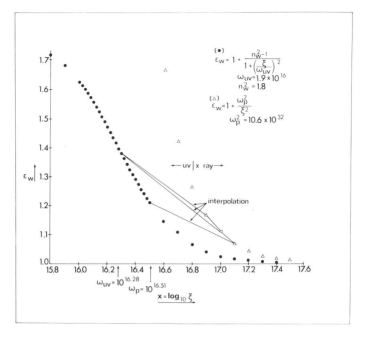

FIG. 3.1. Illustration of uncertainties due to incomplete knowledge of dielectric data in the far ultraviolet. Curves plotted are frequency-dependent dielectric constant of water for the imaginary axis in the ultraviolet to X-ray region. ($\bullet$), Plot of eqn. (3.2); ($\triangle$) plot of eqn. (3.3). Solid lines are examples of possible straight line interpolations. [After Ninham and Parsegian (1970a).]

non-retarded energy of interaction depends roughly on the integral

$$\int_0^\infty \left(\frac{\varepsilon_1 - \varepsilon_2}{\varepsilon_1 + \varepsilon_2}\right)^2 d\xi$$

for two half-spaces "1" separated by "2".) That is, all integrals required involve differences of dielectric functions divided by sums. Since all susceptibilities $\varepsilon(i\xi)$ approach the limiting form (3.3) as $\omega \to \infty$ ($\omega \gg$ ultraviolet frequencies), the integrands of all integrals will tend to zero rapidly beyond the ultraviolet region. Hence if the two media 1 and 2 have similar density their susceptibilities will already be approximately equal at X-ray frequencies. The uncertain region connecting uv with X-ray frequencies then contributes significantly only for those cases where $\varepsilon(\omega)$ describes media of greatly different density. For interactions of organic substances across water or oil all of which have similar weight density the uncertainty introduced by interpolation is much diminished.

We now illustrate this procedure for several substances.* More detailed accounts are given by Ninham and Parsegian (1970a) and by Parsegian and Gingell (1972).

Hydrocarbons

The dielectric susceptibility of a hydrocarbon is known to be essentially constant from zero frequency through to the optical region with a value approximately $\varepsilon \approx 2$. For liquid octane, e.g., the refractive index in the visible is $n = 1.40$; $\varepsilon = n^2 = 1.96$. The complicated and incompletely known ultraviolet absorption spectrum is summarized for the near ultraviolet by taking as a mean value the first ionization potential,[†] for octane $\omega_{uv} = 1.55 \times 10^{16}$ rad/sec (see e.g., "Handbook of Chemistry and Physics"). For octane then we would have up to the ultraviolet

$$\varepsilon(i\xi) = 1 + \frac{(1 \cdot 96 - 1)}{1 + (\xi/1 \cdot 55 \times 10^{16})^2} . \tag{3.4}$$

At frequencies above the plasma frequency we can use (3.3) with ω_p^2 computed from known weight density (0·66 gm/cc), atomic weight (86·2) and electron number/molecule (50). The resulting curves are similar to Fig. 3.1, and some interpolation between say 10^{16} and 10^{17} rad/sec appears necessary.

Water

At 20 °C, $\varepsilon(\omega)$ at zero frequency has the value $\varepsilon(0) \approx 80$ and undergoes the simplest Debye rotational relaxation at $\omega = 1 \cdot 083$ rad/sec, down to a value $\varepsilon(\omega) = 4 \cdot 2$ centred around a value $\omega \approx 1 \cdot 06 \times 10^{11}$. According to

* Useful data can be found e.g., in "The Handbook of Chemistry and Physics," late editions, Chemical Rubber Publishing Co., Cleveland, Ohio; "Tables of Dielectric Data for Pure Liquids and Dilute Solutions," Natl. Bur. Stds (USA); circular **589**, 6 (1958); "Selected Values of the Physical Properties of Hydrocarbons and Related Compounds" (ed. Rossini, Pitzer and Arnett), Am. Petroleum Institute Research Program, Thermodynamics Research Center, College Station, Texas (1967); and in Hirayama (1967). The utility of sources appears to follow a retarded law, being inversely proportional to roughly the cube of its accessibility.

† The assignment of *all* oscillator strengths in the uv to a one-term dispersion equation tends to overestimate $\varepsilon(i\xi)$ in the near ultraviolet. The lumping together of all uv relaxation frequencies into a single frequency taken to be the first ionization potential will usually, for condensed substances, also provide a source of overestimate. One could argue that it would be desirable in such circumstances to introduce no interpolation whatever, because the overestimate of $\varepsilon(i\xi)$ for $\xi < \omega_{uv}$ will be compensated by the neglect of higher relaxation frequencies. Surprisingly this expectation is sometimes borne out by detailed comparison of theory with experiment, but to then claim precise agreement may be pushing the matter too far. The (gaseous) value of ω_{uv} will be different from that in a condensed medium, and on adjustment (usually about 10%) can be made, for example, from the Clausius-Mossotti formula.

Kislovskii (1956, 1957, 1958) who analysed the reflection spectrum of water, there are in the infrared five clearly distinguished peaks. Kislovskii has described these peaks in terms of both the absorption frequencies ω_j, effective indices of refraction far before, and far after the individual absorption bands, and has also given a measure of bandwidths. From this reflection data, we have at least for water good estimates of the parameters γ_j of eqn. (3.2) in the infrared. (For details of the reduction of Kislovskii's data to a convenient form see the paper of Gingell and Parsegian (1972). Their paper is important because they have used this information to study directly and confirm the expectation that calculations are insensitive to bandwidths.) After the infrared, $\varepsilon^{1/2}$ decreases to the measured optical index of refraction $\varepsilon^{1/2} = n = 1\cdot337$. Further strong relaxation occurs in the uv and again we subsume all such frequencies into a single $\omega_{uv} = 1\cdot906 \times 10^{16}$ rad/sec corresponding to the ionization potential, after which available data are insufficient for a more precise description. According to (3.2) then, up to the ultraviolet we should have a representation for $\varepsilon_w(i\xi)$ which consists of a single microwave term, five infrared terms, and one ultraviolet term. Beyond the plasma frequency $\omega_p^2 = 10\cdot6 \times 10^{32}$, the form (3.3) holds, and we must again decide on a suitable interpolation range.

It is of course inconsistent to use precise data in the infrared with less accurate information in the ultraviolet. However, the procedure outlined above has the virtue of simplicity. An alternative method is that developed by Krupp (1967), Krupp, Walter and Schnabel (1972) who use the relation

$$\varepsilon(i\xi) - 1 = \frac{2}{\pi} \int_0^\infty \frac{\omega \varepsilon''(\omega)\, d\omega}{\omega^2 + \xi^2}, \tag{3.5}$$

where $\varepsilon''(\omega)$ is the imaginary part of the complex dielectric constant. $\varepsilon''(\omega)$ is obtained by analysis of reflection measurements, and the integration carried out numerically. In principle this technique although cumbersome should lead to a better numerical representation for $\varepsilon(i\xi)$ than the method outlined above, but reliable data is sparse, and other practical limitations, e.g., the frequency range spanned by the relevant experiment, practically always throw us back to the simpler method. Earlier methods for fitting observable data in the visible and ultraviolet to a form suitable for computational purposes have been reviewed by Gregory (1969), who was concerned only with electronic transitions in the ultraviolet. The procedure most used in the past for calculation of Hamaker constants has been to assume that the molar refraction is related to polarizability by the Lorentz–Lorenz equation

$$\frac{n^2 - 1}{n^2 + 2} \frac{M}{\rho} = \frac{4\pi}{3} N_0 \alpha(\omega), \tag{3.6}$$

where n is the refractive index of the substance at angular frequency ω, M is the molecular weight, ρ the density and N_0 the Avogadro number. Substituting the form

$$\alpha(\omega) = \frac{e^2}{4\pi^2 m} \sum_j \frac{f_j'}{\omega_j^2 - \omega^2}, \qquad (3.7)$$

where f_j' are again oscillator strengths, one has

$$\frac{n^2-1}{n^2+2} \frac{M}{\rho} = \frac{e^2}{3\pi m} \sum_j \frac{f_j'}{\omega_j^2 - \omega^2}. \qquad (3.8)$$

This expression which applies to dielectric materials in either the gaseous or condensed state gives the variation of refractive index with frequency and is one type of dispersion equation.

Equation (3.8) breaks down at frequencies close to strong absorption bands, but for reasons outlined above, the damping terms are usually negligible on the imaginary axis. The standard approach has again been to assume that the variation of refractive index can be represented by an effective dispersion equation with only one term, yielding

$$\frac{n^2-1}{n^2+2} \frac{M}{\rho} = \frac{e^2 N_0}{3\pi m} \frac{s}{\omega_v^2 - \omega^2}, \qquad (3.9)$$

where $s = \sum f_j'$ is regarded as "the effective number of dispersion electrons," and ω_v is an effective characteristic frequency. Plots of $(n^2 + 2)/(n^2 - 1)$ vs. ω^2 for a range of frequencies should give a straight line from which values of s and ω_v can be inferred, and it is these values which have been most used to calculate Hamaker constants. Gregory investigates a number of suggested methods for estimating the "best" values of ω_v. The apparent excellent observed linearity of plots of ω^2 vs. $(n^2 + 2)/(n^2 - 1)$ may be somewhat illusory, since the range of frequencies spanned usually does not go far enough into the ultraviolet. Some interpolation to the near X-ray region is inevitable in the absence of more experimental information. (Precise specification of the frequency ω beyond which the plasma form of $\varepsilon(\omega)$ (eqn. (3.3)) holds is difficult, since the term ω_p^2/ω^2 is the first term in a complicated many-body calculation. But (3.3) should be a reasonable approximation to the actual $\varepsilon(\omega)$ for say $\omega \gtrsim 2\omega_p$.) We mention in passing the paper of Nir, Rein and Weiss (1972) who criticized the procedure used by Ninham and Parsegian (1970b). The paper of Nir, Rein and Weiss is somewhat misleading (see Gingell and Parsegian, 1972).

For metals, the situation is very much more complicated and as yet imperfectly explored. At large distances only the static part of the dielectric constant (here infinite) occurs, but at smaller distances the effects introduced by a conducting medium may require a different approach taking into account properly the spatial dispersion of the dielectric constant which should now be $\varepsilon = \varepsilon(k, \omega)$ where k is the wave number. One difficulty here is that the proper boundary conditions in spatially dispersive media are not known. We return to this problem in Chapter 7.

3.2   Use of dielectric data in comparison with experiment

Since van der Waals forces are involved whenever bodies come into close proximity one can expect to get information on the forces from a variety of phenomena dealing with colloid stability, surface tension and wetting, adhesion, thinning of liquid films and adsorption. But most experiments dealing with such systems determine the forces indirectly, their interpretation often involving many parameters which are themselves uncertain. For example, experiments on the kinetics of flocculation (Ottewill and Shaw, 1966; Watillon and Joseph-Petit, 1966), and on soap film thinning (e.g., Lyklema and Mysels, 1964; Bruil, 1970) require for their interpretation an understanding of the role of electrostatic double layer forces in addition to dispersion forces. [Kinetic experiments are particularly difficult to interpret. Thus the theory of the kinetics of flocculation describes the phenomenon in terms of an integrated force law between the bodies, including electrostatic forces. The latter are sensitive to assumptions made about constant charge, constant potential, charge regulation, and the nature of Stern layers. Besides the theory rests on assumptions of viscosity, shape and other factors.] For the deduction of the van der Waals contribution to interfacial energies from data on surface tension (Israelachvili, 1973; Fowkes, 1968) some information must be assumed on packing of molecules. This is so even if the orientation forces due to permanent dipole moments are absent. Deductions from the strengths of materials (Bailey and Daniels, 1973) or adhesion (van den Tempel, 1972; Krupp, 1967) usually require various assumptions about electrostatic forces or bulk properties. Analysis of virial coefficients, crystal properties and molecular beam experiments necessarily depend for their interpretation on assumptions made about ill-defined short range repulsive forces in the region of overlap of molecular orbitals. An understanding of van der Waals interaction without the complications associated with indirect deductions of the sort mentioned above would obviously be most useful. It is for this reason that some effort has been directed, during the last twenty years or so, towards direct measurement of dispersion forces between solid bodies. The first of such measurements were the famous experiments of Abrikosova and

Deryaguin begun in 1951 who first demonstrated the existence of retarded forces at very large distances, of the order of 1000 Å (Deryaguin and Abrikosova, 1956). The various direct measurement experiments which require enormous ingenuity and skill have been reviewed by Tabor and Israelachvili (1973). The most accurate are those of Israelachvili and Tabor (1972) who measured the force between crossed cylinders of mica, which can be made molecularly smooth. These authors spanned the distance regime from 14 to 1300 Å and convincingly demonstrated the transition from non-retarded to retarded behaviour. The results are in remarkable agreement with the Lifshitz theory. Even more remarkable was their direct demonstration of the effect of material properties on the forces, which we shall discuss further below. Our purpose here is not to duplicate the review of experiments of Tabor and Israelachvili (1973), but to underline the theoretical uncertainties which are unavoidable at close separations in the absence of accurate dielectric data at very high frequencies.

Analysis of experiments on mica plates

The experiments of Israelachvili and Tabor (1972) involved measurement of the force between crossed mica cylinders. In interpreting the results it is therefore necessary to modify the Lifshitz result for planes by insertion of a geometric factor (see §1.4). This assumption holds only in the limit of pairwise additivity, but since the bulk of the interaction comes from ultraviolet frequency correlations and the cylinders have radii very large compared with their separation, it should here be an excellent approximation. If this approximation is made, we require the force across a medium 2, here a vacuum, of width l bounded by two semi-infinite media 1. The result is, from eqns. (2.58) to (2.60)

$$\mathscr{F}(l) = -\frac{k_B T}{2\pi c^3} \sum_{n=0}^{\infty}{}' \xi_n^3 \xi_2^{3/2} \int_1^{\infty} \mathrm{d}p\, p^2 \left(\frac{\exp(-x)}{[1 - \Delta^{-2}\exp(-x)]} \right.$$

$$\left. + \frac{\exp(-x)}{[1 - \bar{\Delta}^{-2}\exp(-x)]} \right), \quad (3.10)$$

where $x = [2lp\xi_n\sqrt{\varepsilon_2}/c]$, and $\Delta, \bar{\Delta}$ are defined by eqns. (2.59), (2.60). The effects of temperature are here completely negligible, so that the sum over imaginary frequencies ξ_n can be replaced by an integral:

$$\sum_{n=0}^{\infty}{}' f(\xi_n) \rightarrow \frac{\hbar}{2\pi k_B T} \int_0^{\infty} \mathrm{d}\xi f(\xi).$$

The change of variable $y = 2l(\sqrt{\varepsilon_2}/c)(p-1)$ then reduces the double integral to the form

$$\mathscr{F}(l) \to \frac{\hbar}{32\pi^2 l^3} \int_0^\infty d\xi e^{-2l/c\sqrt{\varepsilon_2}\xi} \int_0^\infty e^{-y} f(y, \xi) dy, \qquad (3.11)$$

where $f(y, \xi)$ is smoothly varying as a function of y and ξ. In this form a standard Gauss–Laguerre numerical integration routine with a 16-point rule is more than adequate, and the integral can be computed without difficulty. In other situations when temperature is important the sum over ξ_n must be carried out directly. Since it has an effective cut-off built in at $\xi_n > \xi_s$ the sum is always finite. Even for $l \sim 10$ Å at $T \sim 20\,°C$ only about 500 terms of the sum contribute, and no problem exists with round-off. (We remark in passing that in general the approximation of the sum by an integral is not good. Corrections to Euler–Maclaurin or similar summation formulae can be significant depending on the materials involved, are not easily obtained analytically, and can be shown to measure physically a rather complicated interplay between Debye, Keesom and dispersion forces. Hence some care must be exercised in computation and the sum form is preferable.)

We now proceed to calculations relating to mica. Following eqns. (3.2) and (3.3) we use the representation

$$\varepsilon(i\xi) = 1 + \frac{D-R}{1+(\xi/\omega_{mw})} + \frac{R-n^2}{1+(\xi/\omega_{ir})^2} + \frac{n^2-1}{1+(\xi/\omega_{uv})^2}; \qquad \xi \le \omega_3 \quad (3.12)$$

$$\varepsilon(i\xi) = 1 + (\omega_p/\xi)^2; \qquad \xi > \omega_4 \gg \omega_p. \qquad (3.13)$$

Here ω_{mw}, ω_{ir}, ω_{uv} are the major microwave, infrared and ultraviolet absorption frequencies for mica. D and R are the mean values of the susceptibility for zero frequency and for frequencies between microwave and the infrared, and n is the refractive index in the visible region. In the absence of better far-uv data, we are forced to interpolate for frequencies between ω_3 and ω_4 as in §3.1. ω_3 and ω_4 must be chosen so that the susceptibility decreases monotonically with increasing ξ. Data provided by Tabor are $D = 5.4$, $n = 1.5757$, $\omega_{mw} = 1.88 \times 10^{10}$ rad/sec, $\omega_{uv} = 1.57 \times 10^{16}$ rad/sec. Mica contains some water, and in the absence of reliable infrared data we choose $R - n^2 = 2.0$, $\omega_{ir} = 0.6 \times 10^{14}$, an average infrared relaxation frequency. Fortunately the final results are quite insensitive to extreme variations in either R or ω_{ir}. The plasma frequency ω_p evaluated from tabulated weight data is 5.2×10^{16} rad/sec. The results of the calculation are shown in Fig. 3.2 (Richmond and Ninham, 1972). Here curves A and B represent the van der Waals force acting between two plane parallel mica plates as a function of separation l, for different interpolations. The dotted line indicates the slope of the force vs. distance

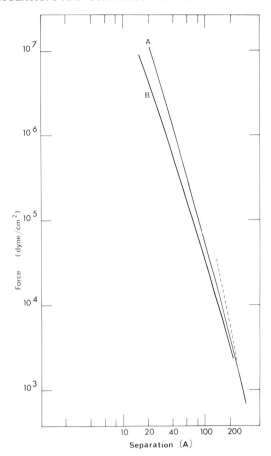

FIG. 3.2. Calculated values of van der Waals force between two plane parallel mica plates as a function of separation l for different interpolations of mica susceptibility in the ultraviolet. [Taken from Richmond and Ninham (1972).]

curve in the fully retarded region. Curve A is a plot of the computed van der Waals force against plate separation for an interpolation $\omega_3 = 1 \cdot 57 \times 10^{17}$ rad/sec, $\omega_4 = 10^{19}$ rad/sec. Curve B is a similar plot without interpolation, i.e., all oscillator strengths are assigned to the closest relaxation frequency. Interpolation introduces uncertainties only in the short distance regime $l \lesssim 60$ Å. The curves gradually change from non-retarded (slope 2) to retarded behaviour as separation increases. For $l \gtrsim 60$ Å experimental points lie precisely on the computed curves. In the retarded regime we expect $\mathscr{F} \sim -B/l^4$. The experimental result is $B = (0 \cdot 97 \pm 0 \cdot 06) \times 10^{-19}$ erg cm. The theoretical result is $B = 0 \cdot 93 \times 10^{-19}$, so that theory and experiment are in complete agreement. In the non-retarded region Israelachvili and Tabor, writing $\mathcal{F} = A/l^3$, find $A = (1\cdot35 \pm 0\cdot15) \times 10^{-12}$ erg. Calculations yield (curve A) $A \approx 1\cdot4 \times 10^{-12}$ erg, and (curve B) $A = 0\cdot7 \times 10^{-12}$ erg (at $l = 20$ Å). The last is almost certainly a lower bound to the true A value because of the absence of interpolation which underestimates high frequency contributions which dominate at close distances. Without more refined data in the uv we can do no better.

Not content with one experiment of singular ingenuity, Israelachvili and Tabor (1972) have gone further, and studied the effect of adsorbed layers. They were able to deposit stearic acid monolayers of precisely determined thickness 25 Å on each surface of their interacting cylinders. The interaction is now between triple layers for which (Ninham and Parsegian, 1970a) the energy of interaction has the non-retarded form

$$E(l, b) \approx -\frac{\hbar}{16\pi^2} \int_0^\infty \left(\frac{\Delta_{32}^2}{l^2} + \frac{\Delta_{21}^2}{(l+2b)^2} + \frac{2\Delta_{32}\Delta_{21}}{(l+b)^2} \right) d\xi. \quad (3.14)$$

In this expression "1" represents mica, "2" the adsorbed layers of stearic acid of thickness b and "3" the intervening medium, here a vacuum of width l. For $l \gg b$, this approaches the energy of interaction of two mica plates

$$E(l, b)_{l \to \infty} \to -\frac{\hbar}{16\pi^2 l^2} \int_0^\infty \Delta_{32}^2 d\xi. \quad (3.15)$$

For $l \leq b$, the energy and force are dominated by the properties of the adsorbed layer—the "effective" Hamaker "constant" depends on distance. By assigning bulk properties to the adsorbed layer, the experiment confirmed this prediction. At distances l greater than 50 Å the measured force tended towards the theoretical value for mica, while for separations less than 25 Å tended towards the value for stearic acid. This result confirms in a direct and most satisfying manner that material bulk dielectric properties are indeed of central importance in determining interactions. The effect of adsorbed coats on van der Waals interactions between spheres has also been studied by Langbein (1969, 1971).

A number of other excellent experiments on direct measurement of van der Waals forces between macroscopic bodies have been carried out. All confirm the correctness of Lifshitz theory. For a discussion of these experiments, and for references to work not cited here, see Kitchener and Prosser (1957); Black, de Jongh, Overbeek and Sparnaay (1960); van Silfhout (1966); Tabor and Winterton (1969); Rouweler and Overbeek (1970); Wittmann, Splittgerber and Ebert (1971); Hunklinger, Geisselmann and Arnold (1972); Bargeman (1972); and van Voorst Vader (1972).

The experiment of Anderson and Sabisky

An experiment of equal precision is that of Anderson and Sabisky (1970, 1973) who measured the thickness vs. height of thin liquid helium films on a vertical substrate. Earlier studies, like those involving direct measurements of van der Waals forces were hampered by the difficulty of obtaining a clean substrate on which the film forms and other complications. With a judicious choice of substrate, such investigation can provide a particularly clean experimental system. Anderson and Sabisky used cleaved surfaces of alkaline earth fluorides for which dielectric data is available, and which are free from the complications for metals.

For a vertical saturated film in equilibrium with its vapour it can be shown (Dzyaloshinskii, Lifshitz and Pitaevskii, 1961) that the thickness l of the film is related to the height z above the bulk liquid by

$$z(l) = -\frac{[\partial F(l)/\partial l]}{\rho g}. \qquad (3.16)$$

Here ρ is the liquid mass density, g the acceleration due to gravity, and $F(l)$ the free energy of interaction of the substrate with air across a planar slab of liquid helium. Since $F(l) \propto 1/l^2$ (non-retarded) and $\propto 1/l^3$ (retarded) we expect the film height to behave as $z \propto 1/l^3$ for small thickness l, and as $1/l^4$ in the retarded region. The third power law was predicted theoretically many years ago, but the experiments have been extremely difficult to carry out. Final confirmation of the third power law had emerged by the late 1950s (Anderson, Liebenberg and Dillinger, 1960), but definitive comparison of theory and experiment had to wait until 1972. Details are given by Sabisky and Anderson (1973) and theoretical calculations for the system by Richmond and Ninham (1971). The results are illustrated in Fig. 3.3. Both retarded and non-retarded behaviour are clearly apparent. Three different interpolations are shown by the curves. Interpolation makes essentially no difference for $l \gtrsim 50$ Å when retardation sets in and agreement with experiment is excellent. The apparent precise agreement for the lower curve (no interpolation) is probably fortuitous. For further discussion of this experiment and for further theoretical calculations see the excellent paper of Sabisky and Anderson (1973).

Notice again that the absence of far ultraviolet data leaves some uncertainty in the non-retarded region $l \lesssim 50$ Å. One additional feature of the experiments of Israelachvili and of Sabisky and Anderson deserves note. If we assign all absorption in the visible and ultraviolet to a single ultraviolet peak centred at the ionization potential with oscillator strength given by the refractive index in the visible, agreement with experiment is

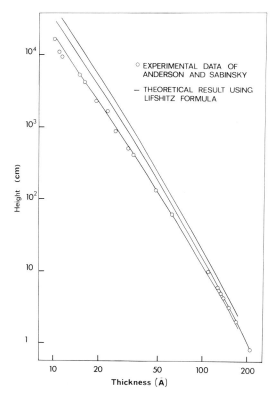

FIG. 3.3. Film thickness vs. height for helium on a vertical BaF$_2$ substrate. [Taken from Richmond and Ninham (1971).]

obtained which is almost too good to be believable, i.e., interpolation due to incomplete knowledge of far uv spectra can be forgotten. This is probably a reflection of the circumstance that the materials involved in these experiments do have very simple absorption spectra. In general this will not be so.

3.3 Condensed media interactions

As emphasized above uncertainties which remain in comparing theory with experiment in the limit of very close separation have more to do with unavailability of reliable ultraviolet dielectric data than anything else. Direct measurements involving interactions across or with a vacuum emphasize high frequency contributions where data is sparse. For condensed media interactions between substances of similar weight density, as already mentioned in §3.1, the relative importance of ultraviolet correlations is much diminished and we can expect more interesting information to emerge which is particularly relevant to organic materials and biological substances. In this regard a most important experiment was that of Haydon and Taylor (1968), who measured the contact angle of a thin non-aqueous film in water. The work has been carried further by Haydon and his collaborators, and is so important that his results deserve a full theoretical investigation. The first film was of n-decane in water stabilized by glycerol monoleate, and its thickness was close to twice the chain length of the stabilizer molecules, indicating that steric repulsion was preventing further thinning. On the assumption that the depth of the free energy minimum is entirely due to van der Waals interactions they were able to deduce that the "effective Hamaker constant" for a film of thickness 50 Å was $4 \cdot 7 \times 10^{-14}$ erg. Difficulties in determining the thickness by a capacitance method probably restrict the accuracy of the result to within a maximum of about 50%. The corresponding value deduced by pairwise summation methods, i.e., assuming all contributions come from the ultraviolet correlations is about $0 \cdot 9 \times 10^{-14}$ erg, a striking discrepancy, which can be explained by using the Lifshitz theory (Parsegian and Ninham, 1969, 1970; Ninham and Parsegian, 1970b; Parsegian and Gingell, 1972).

We first introduce some notation. Again from eqn. (2.58) we have, denoting water as medium "1," and the intervening hydrocarbon as "2,"

$$F(l, T) = \frac{k_B T}{8 \pi l^2} \sum_{n=0}^{\infty}{}' I(\xi_n, l), \qquad (3.17)$$

where $I(\xi_n, l)$ is defined in eqn. (2.59), and $F(l, T)$ is the free energy of interaction. We define

$$F(l, T) = -\frac{A(l, T)}{12 \pi l^2}, \qquad (3.18)$$

where $A(l, T)$ is now the "Hamaker function." Comparing eqn. (3.18) with (3.17) we have

$$A(l, T) = -\tfrac{3}{2} k_B T \sum_{n=0}^{\infty} I(\xi_n, l). \qquad (3.19)$$

At low temperature

$$A(l, 0) \rightarrow -\frac{3\hbar}{4\pi} \int_0^{\infty} I(\xi, l) \, d\xi \qquad (3.20)$$

and in the limit $l \to 0$ (no retardation)

$$A(0,0) = -\frac{3\hbar}{4\pi} \int_0^\infty \mathrm{d}\xi \int_0^\infty x \, \mathrm{d}x \, \ln\left\{1 - \left(\frac{\varepsilon_2 - \varepsilon_1}{\varepsilon_2 + \varepsilon_1}\right)^2 \mathrm{e}^{-x}\right\}$$

$$\approx \frac{3\hbar}{4\pi} \int_0^\infty \mathrm{d}\xi \left(\frac{\varepsilon_2 - \varepsilon_1}{\varepsilon_2 + \varepsilon_1}\right)^2. \quad (3.21)$$

Unfortunately the dielectric properties of the film are not known, but all hydrocarbons have a dielectric response similar to that given in eqn. (3.4). Using a set of dielectric values for hydrocarbon which covers the extreme limits for hydrocarbons, Ninham and Parsegian (1970b) and Parsegian and Ninham (1970) have computed the Hamaker function using the techniques outlined above. These calculations were refined using more accurate infrared absorption data for water by Gingell and Parsegian (1972). The results are given in Table 3.1 below. The value 1·9 for the square of the refractive index is certainly too low, and for Haydon's film the correct value probably lies between 2·0 and 2.1. Note that the agreement with the measured value $4·7 \times 10^{-14}$ erg is excellent. The effect of uncertainties in far uv spectra are very small because of (1) retardation which acts to remove very high frequency contributions and (2) the circumstances that the materials have similar weight density and electron densities. Detailed studies of the effects of variations of the computed results to choice of ionization potential, and of the influence of band-widths in the infrared have been made by Ninham and Parsegian (1970b) and by Gingell and Parsegian (1972).

Spectral contributions

It is of considerable interest to examine the dependence of the disper-sion energy on various frequency regions. To do this consider the integral approximation (3.20). $I(\xi, l)$ as defined by eqn. (2.58) measures the contribution to the energy of frequencies in the range $(\xi, \xi + \mathrm{d}\xi)$. Chang-ing variables to $x = \log_{10}(\xi)$ the relative contribution from the spectral

Table 3.1. Estimates of $A(l, T)$ using Lifshitz expression

n_{hc}^2	A(erg) $\omega_{uv}^{hc} = 1\cdot54 \times 10^{16}$ (1-decane ionization potential)	A(erg) $\omega_{uv}^{hc} = 1\cdot76 \times 10^{16}$ (ethane ionization potential)
1·9 (n-hexane)	3.44	3.51
2·0	3.82	4.23
2·1	4·59	5·35
2·2	5·70	6.85

Data for water as in §3.1, $l = 50$ Å, $T = 20$ °C.

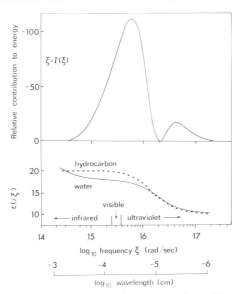

FIG. 3.4. Relative contribution to the Hamaker constant from different frequency regions for the hydrocarbon–water system. [After Gingell and Parsegian (1972).]

range $(x, x + dx)$ is then $2\cdot303\xi I(\xi, l)\,dx$. Figure 3.4 shows plots of $I(\xi, l = 0)$ for assumed hydrocarbon values $n_{hc}^2 = 2\cdot0$, $\omega_{uv}^{hc} = 1\cdot76 \times 10^{16}$ rad/sec. The Hamaker constant is roughly proportional to the total area under the curve plus an added zero frequency term, which for this system is very large. The upper curve shows relative contributions to interaction energy of water across a hydrocarbon film at different frequencies. The lower curve illustrates the source of interaction energy in regions where dielectric susceptibilities of water and hydrocarbon differ. About 80% of the non-zero frequency interaction energy results from visible and ultraviolet fluctuations, and about 20% from the infrared. The possibility that infrared fluctuations, and for that matter visible frequency fluctuations, could be significant in van der Waals interactions had not been seriously considered in the earlier literature. The relative contribution of different frequency regimes can be a sensitive function of refractive indices, and of separation. To illustrate this we plot in Fig. 3.5 a similar curve to that of Fig. 3.4 but with different (model) infrared data for water (Ninham and Parsegian, 1970b). This data is incorrect, and overestimates the contribution of the infrared at the expense of the visible. Better data is used by Gingell and Parsegian (1972), but Fig. 3.5 illustrates the point. The figure illustrates the progressive removal of high frequency contributions with increasing l. Infrared contributions are not damped until $l \sim 500$ Å, while for ultraviolet fluctuations begin to damp out at $l \sim 50$ Å.

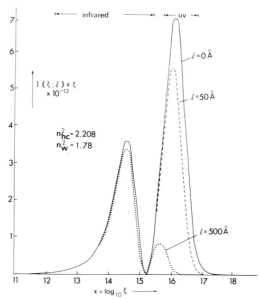

FIG. 3.5. Model calculation illustrating the influence of retardation on relative contributions from different frequencies to van der Waals interactions. [From Ninham and Parsegian (1970b).]

For interactions of hydrocarbon across water, the role of the infrared is not so remarkable. Generally, when water is involved with organic materials—and water has a very strong infrared adsorption—one can expect a peculiar and delicate interplay between relative frequency contributions and retardation. In some situations, the interaction between unlike materials can be attractive at large distances, turn repulsive at smaller distances and again turn attractive at even smaller distances! (Richmond, Ninham and Ottewill, 1973; Parsegian and Gingell, 1973.) Experimental confirmation of this theoretical prediction comes from simple studies of the spreading of hydrocarbon films on water, to be discussed in §3.5.

3.4 Temperature-dependent forces*

Returning to (3.21) we now compute the Hamaker constant $A(0, 0)$ for the Haydon film.

* There is a caveat to this section. While the conclusions and nature of the force are correct for salt-free water, in the presence of ions the nature of the temperature-dependent forces changes completely at distances beyond the Debye screening length of the ions in a complicated manner which depends also on geometry. For planar interactions of water with water across another (salt-free) substance the effect of this screening is negligible. In the opposite situation it is. This will be partially studied in Chapter 7.

Depending on the refractive index chosen to model the hydrocarbon, (cf. Table 3.1), the Hamaker constant computed from eqn. (3.21) takes a value lying between approximately $0 \cdot 2 \times 10^{-14}$ erg $(n_{hc}^2 = 1 \cdot 9)$ up to a maximum of $2 \cdot 3 \times 10^{-14}$ erg $(n_{hc}^2 = 2 \cdot 1)$. For the most likely value of n_{hc} the contribution of ultraviolet, visible and infrared frequencies is $\approx 10^{-14}$ erg. A similar result emerges if we compute these contributions directly from the sum (3.19), omitting terms in $\xi_n = 0$. But for oil–water systems, the measured and theoretical Hamaker constant is $A(50 \text{ Å}, T = 20 °C) \approx 4 \cdot 7 \times 10^{-14}$ erg $\approx k_B T > 10^{-14}$ erg as it would be if ultraviolet and infrared contributions dominate. For these and similar systems then a good part of the whole interaction comes from the zero frequency of the sum (3.17). This term is proportional to and increases almost linearly with temperature. Consequently, being independent of $\hbar$ it has to do with molecular correlations or orientation forces between the permanent dipolar molecules in liquid water. We now investigate this contribution* in some detail following Parsegian and Ninham (1970). From (3.17) and (3.18) we may write

$$A(l, T) = A_{n=0} + A_{n>0}. \tag{3.22}$$

At room temperature $T \approx 293 °K$, the frequencies ξ_n at which the terms of the sum of (3.17) are to be evaluated are $\xi_n = 2 \cdot 41 \times 10^{14}$ rad/sec, $n = 0, 1, 2, \ldots$. To an approximation good for our present purposes terms in the sum beyond that in $n = 0$ are closely spaced and can be replaced by an integral

$$A_{n>0} \sim -\frac{3\hbar}{4\pi} \int_{2\pi k_B T/\hbar}^{\infty} I(\xi, l) \, d\xi$$

which is practically the same as that already considered. The term in $n = 0$ uses zero frequency dielectric constants and from (3.17), (3.19) and (2.59) is explicitly

$$A_{n=0} = -\tfrac{3}{4}k_B T I(0, l) = -\tfrac{3}{4}k_B T \int_0^{\infty} x \, dx \ln (1 - [\Delta_{12}(0)]^2 e^{-x})$$

$$= \tfrac{3}{4}k_B T \sum_{j=1}^{\infty} \frac{\Delta_{12}^{2j}(0)}{j^3}, \tag{3.23}$$

---

*There is much confusion in the literature about the relevance or otherwise of temperature-dependent forces. While Lifshitz in his 1956 paper appears to have suspected that temperature forces could be important, subsequent Russian literature, e.g., Abrikosov, Gor'kov and Dzyaloshinskii (1965), asserts that temperature-dependent contributions are *always* negligible for film thicknesses $l \ll 0$ $(\hbar c/k_B T) \approx 10^5$ Å at room temperature. The inequality is a necessary condition for the neglect of the temperature-dependent orientation forces, but it is not sufficient.

where

$$\Delta_{12}(0) = \frac{\varepsilon_2(0) - \varepsilon_1(0)}{\varepsilon_2(0) + \varepsilon_1(0)}. \tag{3.24}$$

At $k_B T = 4 \times 10^{-14}$ erg we have

$$A_{n=0} \approx 3 \cdot 2 \times 10^{-14} \text{ erg}. \tag{3.25}$$

This is more than 50% of Haydon's measured value. Not only is the contribution $A_{n=0}$ similar in magnitude to the sum of higher frequency terms, but it is an infinite wavelength contribution. This circumstance assures that in oil–water systems the van der Waals energy will have a short distance or non-retarded form even at very large distances, and is completely contrary to the usual intuition regarding the value of van der Waals forces.

Low density limit

In the low density limit, the contribution due to the $n = 0$ term reduces to the Keesom force acting between permanent dipolar molecules. To see this, consider two clouds of water molecules interacting across a vacuum, a planar gap of width $l \ll N^{-1/3}$ where N is the number density of water molecules. For water vapour (Fröhlich, 1958) we take

$$\varepsilon_w(0) = 1 + 4\pi\mu^2 N/3k_B T, \tag{3.26}$$

where μ is the dipole moment of an individual molecule. From (3.17)–(3.19), (3.23), we have then

$$F_{n=0} = -\frac{k_B T}{16\pi l^2}\left(\frac{\varepsilon_w(0) - 1}{\varepsilon_w(0) + 1}\right)^2 \approx -\frac{\pi}{36 l^2}\frac{\mu^4 N^2}{k_B T}. \tag{3.27}$$

The entropy associated with this free energy is

$$S = -\left.\frac{\partial F}{\partial T}\right|_P = \frac{F}{T}. \tag{3.28}$$

Hence the enthalpy, or equivalently energy here, is

$$H \equiv E = F + TS = 2F.$$

This is precisely what one obtains by carrying out a pairwise addition of the Keesom intermolecular interaction energies

$$V_{\text{Keesom}}(r) = -\frac{2\mu^4}{3k_B T r^6} \tag{3.29}$$

acting between two permanent dipoles in this configuration. For interactions across a dilute intermediate substance consisting of non-polar molecules, one can substitute $\varepsilon_2 = 1 + 4\pi\rho_2\alpha_2^2(i\xi)$ for the dielectric properties of the intervening medium. Repetition of the process then shows that the $\cdot n = 0$ term includes also the Debye force due to dipole-induced dipole interactions. Corresponding to eqn. (3.27) we now have

$$F_{n=0} \approx -\frac{k_B T\pi}{4l^2}\left\{\left(\frac{\mu^2 N}{3k_B T}\right)^2 + \rho_2^2\alpha_2^2(0) - \frac{2\mu^2\rho_2 N\alpha_2(0)}{3k_B T}\right\} \tag{3.30}$$

and

$$E = F + TS = -\frac{\pi\mu^4 N^2}{18l^2 k_B T} + \frac{\pi\rho_2 N\mu^2\alpha_2(0)}{6}\frac{1}{l^2}. \tag{3.31}$$

This represents the sum of Keesom interaction energies diminished by the sum of permanent dipole-induced dipole forces determined by a potential

$$V_{\text{Debye}} = -\frac{\mu^2\alpha_2(0)}{r^6}.$$

We note especially that the temperature-dependence of the intermolecular force as predicted by pairwise summation is completely wrong for condensed media interactions, and totally misleading.

An additional feature of the temperature-dependent contribution to the forces deserves note. The term $n = 0$ for condensed oil–water systems is primarily an entropic contribution to the free energy. To see this recall that the free energy change in bringing two hydrocarbon bodies separated by water from infinite separation to a distance l is to leading order

$$F_{n=0} \approx -\frac{k_B T}{16\pi l^2}\left(\frac{\varepsilon_w(0) - \varepsilon_{hc}(0)}{\varepsilon_w(0) + \varepsilon_{hc}(0)}\right)^2. \tag{3.32}$$

Here as before the ε's refer to static dielectric constants. Since

$$F_{n=0} = H_{n=0} - TS_{n=0}, \tag{3.33}$$

the entropy change corresponding to this free energy change is

$$-TS_{n=0} = T\frac{\partial F_{n=0}}{\partial T} = -\frac{k_B T}{16\pi l^2}\Delta_0^2\left(1 - \frac{2T}{\Delta_0}\frac{\partial\Delta_0}{\partial T}\right)$$

$$= F_{n=0}\left(1 - \frac{2T}{\Delta_0}\frac{\partial\Delta_0}{\partial T}\right), \tag{3.34}$$

where

$$\Delta_0 = (\varepsilon_w(0) - \varepsilon_{hc}(0))/(\varepsilon_w(0) + \varepsilon_{hc}(0)).$$

Similarly the enthalpy change is

$$H_{n=0} = -T^2\frac{\partial(F/T)}{\partial T} = -F_{n=0}\frac{2T}{\Delta_0}\frac{\partial\Delta_0}{\partial T}. \tag{3.35}$$

Now at 20 °C,

$$\varepsilon_w(0) \approx 80, \qquad \varepsilon_{hc}(0) \approx 2, \qquad \Delta_0 \approx 0.95 \qquad (3.36)$$

and from handbook data we have

$$\frac{\partial \varepsilon_{hc}(0)}{\partial T} \approx -0.016, \qquad \frac{\partial \varepsilon_w(0)}{\partial T} \approx -0.37 \qquad (3.37)$$

whence

$$\frac{\partial \Delta_0}{\partial T} \approx -1.8 \times 10^{-4}.$$

Then from (3.34) we have

$$-TS_{n=0} \approx F_{n=0}[1 - 0.11] \approx 0.9 F_{n=0}. \qquad (3.38)$$

This equation implies that the entropy increases upon approach of the hydrocarbon bodies, and that the free energy change is almost all due to an entropy change for the $n = 0$ contribution. The enthalpy contribution for this case is relatively small

$$H_{n=0} \approx 11\% \ F$$

but has the same sign as the free energy change.

Macromolecule interactions in solution

It had been argued many years ago (Kirkwood and Shumaker, 1952) that proton fluctuations in the microwave region can correlate to produce an effective attractive force between say protein molecules, which carry on their surfaces many ionizable groups. In fact any induced fluctuation in orientation or distortion of a molecule can cause local electromagnetic fluctuations and contribute to an attractive force between like molecules. The magnitude of such forces can be estimated by the Lifschitz formalism, for the effect of these electromagnetic fluctuations will be an increase in the effective dielectric constant $\varepsilon_p(0)$ of a protein solution. To calculate these microwave frequency contributions to the forces we can proceed as follows. Imagine a dilute solution of protein molecules. Then through the full frequency range the dielectric susceptibility may be written as

$$\varepsilon_p(i\xi) = \varepsilon_w(i\xi) + N \frac{\partial \varepsilon_p(i\xi)}{\partial N}, \qquad (3.39)$$

where the subscripts p and w denote protein solution and water respectively, while N is the number density of protein molecules. Consider now the interaction of protein molecules acting between two solutions separated by a gap l containing pure water. Regarding medium 1 as protein solution and medium 2 as water, the free energy of interaction is

$$F(l, T) = \frac{k_B T}{8 \pi l^2} \sum_{n=0}^{\infty}{}' \int_0^{\infty} x \, dx \ln \left\{ 1 - \left(\frac{N}{2 \varepsilon_w} \frac{\partial \varepsilon_p / \varepsilon_w}{\partial N} \right)^2 e^{-x} \right\}. \tag{3.40}$$

To obtain this expression we have ignored retardation and replaced $(\varepsilon_p - \varepsilon_w)/(\varepsilon_p + \varepsilon_w)$ by $(N \partial \varepsilon_p / \partial N)/2 \varepsilon_w$. This approximation is permissible since for a dilute solution $(N \partial \varepsilon_p / \partial N) \ll \varepsilon_w$. Keeping only the leading term in (3.40) we have

$$F(l, T) \approx - \frac{k_B T}{8 \pi l^2} N^2 \sum_{n=0}^{\infty}{}' \left\{ \left(\frac{\partial \varepsilon_p (i \xi_n)}{\partial N} \right) \Big/ \varepsilon_w (i \xi_n) \right\}^2. \tag{3.41}$$

This is precisely what we would have obtained by a pairwise summation of individual protein–protein free energies of interaction of the form

$$f(r) = - \frac{3 k_B T}{8 \pi^2 r^6} \sum_{n=0}^{\infty}{}' \left(\frac{\partial \varepsilon_p (i \xi_n)}{\partial N} \Big/ \varepsilon_w (i \xi_n) \right)^2. \tag{3.42}$$

As in our previous analysis we decompose this free energy of interaction into two terms and write

$$f(r) \approx - \frac{3}{16 \pi^2 r^6} \left\{ k_B T \left(\frac{\partial \varepsilon_p (0)}{\partial N} \Big/ \varepsilon_w (0) \right)^2 \right.$$

$$\left. + \frac{4 \hbar}{\pi} \int_{2 \pi k_B T / \hbar}^{\infty} \left(\frac{\partial \varepsilon_p (i \xi)}{\partial N} \Big/ \varepsilon_w (i \xi) \right)^2 d \xi \right\}. \tag{3.43}$$

The first term concerns relaxation in the microwave region, while the second accounts for fluctuations at infrared and uv frequencies. The error in replacing the second term by an integral rather than a sum over frequencies is not serious for purposes of estimation. It is tantamount to ignoring the absorption of protein solution in the near infrared.

We can now compare the magnitude of these two terms using data for the dielectric increment $\partial \varepsilon_p (0) / \partial \bar{\rho}$ at zero frequency where $\bar{\rho}$ is weight density of protein in solution, and for index of refraction increment $(\partial n / \partial \bar{\rho})$ at optical frequencies. For example, for haemoglobin (Parsegian, 1970) we have

$$\frac{\partial \varepsilon_p (0)}{\partial \bar{\rho}} \sim 0 \cdot 3 \text{ litre/gm} \tag{3.44}$$

and for most proteins in the visible range

$$\frac{\partial \varepsilon_p (i \xi)}{\partial \bar{\rho}} \approx 2 n_w \frac{\partial n_p}{\partial \bar{\rho}}; \qquad \frac{\partial n_p}{\partial \bar{\rho}} = 1 \cdot 8 \times 10^{-4} \text{ litre/gm}. \tag{3.45}$$

In order of magnitude the first term in the curly bracket of (3.43) is $k_B T(\frac{3}{80})^2 \approx 5\cdot 6 \times 10^{-19}$ apart from the common scaling factor $(\bar{\rho}/\rho)^2$. The second can be evaluated if we assume that the principal uv-absorption of protein is roughly at $\xi \gtrsim 10^{16}$ rad/sec. Thus

$$\frac{4\hbar}{\pi} \int_{2\pi k_B T/\hbar}^{\infty} \left(\frac{\partial \varepsilon_p(i\xi)}{\partial \bar{\rho}} \middle/ \varepsilon_w(i\xi) \right)^2 d\xi \approx \frac{4\hbar}{\pi} \int_0^{10^{16}} d\xi$$

$$\times \frac{4(1\cdot 8)^2 \times 10^{-8}}{n_w^2} \sim 5 \times 10^{-19} \quad (3.46)$$

so that for the above data low frequency (microwave) contributions to the protein–protein interaction are of the same order of magnitude as high frequency van der Waals forces in the short distance regime. To the extent that the dielectric increment is due to proton fluctuations, the microwave contribution is due to the Kirkwood and Shumaker (1952) forces. For highly polar protein molecules the dielectric increment could easily increase by a considerable factor, say 10, to make the microwave contribution dominant by one or two orders of magnitude. The important observation however, is that the magnitude and selectivity of van der Waals forces acting between macromolecules should be accessible through macroscopic measurements which are model independent. The necessary modifications of the above argument due originally to Pitaevskii (1959) required to take into account interaction between dissimilar macromolecules and of the effects of solvents are clear. We make two final remarks (i) the caveat to this section must be kept in mind; and (ii) simple arguments like those of the preceding paragraph break down at separations of the molecules of the order of their size. At such close distances the effects of geometry are obviously all important.

3.5 Hydrocarbon films on water

Before leaving our comparison of theory with experiment, we consider one further experiment, the spreading of hydrocarbons on water (Hauxwell and Ottewill, 1970). Five different normal hydrocarbons were considered (pentane, heptane, hexane, octane and dodecane) and it was found that the first three formed a stable wetting [macroscopic] film on a pure water surface, while the last two do not. These hydrocarbon liquids are very similar and have simple dielectric properties which can be modelled with only a single ultraviolet absorption frequency. As can be seen from Table 3.2 the differences between their properties are very small indeed, and since no complications due to electrostatic effects are present, the

Table 3.2. Dielectric properties of paraffins

	n (refractive index)	ω_{uv} (rad/sec) (1st Ionization potential)	ρ (gm/cc)	Atomic weight	Electron number
Pentane	1·36	$1·57 \times 10^{16}$	0·62	72	42
Hexane	1·37	1·55	0·66	86	50
Heptane	1·39	1·50	0·68	100	58
Octane	1·40	1·49	0·70	114	66
Dodecane	1·43	1·40	0·75	170	98

phenomenon is evidently one of some delicacy which provides a challenging test of the Lifshitz theory. The theory of thin films has been developed by, e.g., Landau and Lifshitz (1958), and by Dzyaloshinskii, Lifshitz and Pitaevskii (1961). The latter authors give a particularly lucid qualitative discussion of the constraints on the chemical potential of a film which must hold in order that it wet a supporting surface. A variety of different situations are theoretically possible, giving rise to stable wetting, metastable and unstable wetting films. We briefly summarize the main points. Consider a planar film of thickness l lying on the surface of a solid body 1, and in thermodynamic equilibrium with its vapour 2, at pressure p. If we suppose the liquid is incompressible and neglect the dependence of the bulk chemical potential $\mu_L^{(0)}$ on pressure, the chemical potential per unit volume of the film

$$\mu_L^{(f)}(p, T) = \mu_L^{(0)}(T) + \frac{\mu_{vw}(l, T)}{N}; \qquad \mu_{vw}(l, T) = \frac{\partial F(l, T)}{\partial l}, \quad (3.47)$$

where $\mu_L^{(f)}(p, T)$ is the chemical potential of the film, $\mu_{vw}(l, T)$ is the "thin film" part of the chemical potential due to van der Waals interactions, and N is the number of hydrocarbon molecules per unit volume.* Further, assuming the vapour to be an ideal gas, we can deduce the chemical potential $\mu_V(p, T)$ of the gas phase from the relation

$$\mu_V(p, T) = \mu_V^{(0)}(p_{sat}, T) + k_B T \ln (p/p_{sat}), \qquad (3.48)$$

* This follows from the observation that the chemical potential, being the derivative of the free energy with respect to the number of particles, is determined by the increase in the free energy of the film when an infinitely thin layer of hydrocarbon liquid is added. If the pressure is sufficiently close to the saturated vapour pressure, we can neglect the heat of adsorption of thin layer from the gas phase (cf. §6.1). Then the difference between $\mu_L^{(f)}$ and $\mu_L^{(0)}$ will depend only on the difference between the interaction energies between the molecules of this infinitely thin layer and the water substrate, and the interaction which would occur between these same molecules and hydrocarbon molecules filling the volume occupied by the water.

where p_{sat} is the saturated vapour pressure. At equilibrium the chemical potentials of the vapour and film must be equal, whence

$$\mu_{vw}(l, T) = k_B T N \ln (p/p_{sat}). \tag{3.49}$$

Inversion of this equation yields the adsorption isotherm, i.e., we can exhibit p/p_{sat} explicitly as a function of l. (If the film is under gravity on a vertical wall, then $p = p_{sat} \exp(-mgz/k_B T)$, where m is the mass of a molecule and z is the height above the level of the bulk liquid, and we recover eqn. (3.16):

$$\mu_{vw}(l, T) + \rho g z = 0.)$$

Note that from eqn. (3.49), since $p/p_{sat} < 1$, $\ln p/p_{sat} < 0$, and the chemical potential must be negative. This is a physically sensible requirement as can be seen from Fig. 3.6. The van der Waals force per unit area across the film (usually denoted by $-\pi_D$, the disjoining pressure)

$$\mathcal{F} = -\frac{\partial F(l, T)}{\partial l} = -\mu_{vw}(l, T)$$

must act in the direction shown to balance the pressure due to the vapour phase. The remaining requirement for thermodynamic stability is (Landau

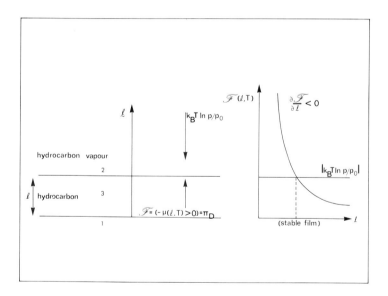

FIG. 3.6. Balance of forces for stable wetting films.

and Lifshitz, 1958)

$$\frac{\partial \mu_{vw}(l, T)}{\partial l} > 0, \quad \text{or} \quad \frac{\partial \mathcal{F}}{\partial l} < 0. \tag{3.50}$$

Again this is a fairly obvious requirement (cf. Fig. 3.6).

We consider now the hydrocarbon–water system in detail (Richmond, Ninham and Ottewill, 1973). From eqn. (2.58) the general formula for the chemical potential required is

$$\mu_{vw}(l, T) = \frac{k_B T}{\pi c^3} \sum_{n=0}^{\infty}{}' \xi_n^3 \varepsilon_3^{3/2} \int_1^{\infty} dp \, p^2$$

$$\times \left(\frac{1}{(\exp\{p\xi_n \sqrt{\varepsilon_3} 2l/c\}/\Delta_{13}\Delta_{25} - 1)} + \Delta \rightarrow \bar{\Delta} \right), \tag{3.51}$$

where the symbols have their usual meanings. If retardation is ignored, then we have the limiting form

$$\mu_{vw}(l, T) \underset{l \to 0}{=} \frac{k_B T}{4\pi l^3} \sum_{n=0}^{\infty}{}' \left(\frac{\varepsilon_1 - \varepsilon_3}{\varepsilon_1 + \varepsilon_3} \right) \left(\frac{\varepsilon_2 - \varepsilon_3}{\varepsilon_2 + \varepsilon_3} \right). \tag{3.52}$$

The second factor in brackets is always negative for all imaginary frequencies $i\xi_n$ since $\varepsilon_2 = 1$ (vapour) and $\varepsilon_3 > 1$. The first factor in brackets depends on the magnitudes of the susceptibilities of water and hydrocarbon liquid and can be either positive or negative. It is convenient to consider three frequency regions:

(1) $\xi < \xi_{ir}$, where ξ_{ir} is a characteristic infrared relaxation frequency for water. (The hydrocarbons have no infrared relaxation.) In this region $80 > \varepsilon_1 > \varepsilon_3 \approx 2$. Thus the contribution to μ is negative.

(2) $\xi_{ir} < \xi < \xi_{uv}(hc)$, where $\xi_{uv}(hc)$ is a characteristic paraffin ultraviolet relaxation frequency. ($\xi_{uv}(hc) \approx 1 \cdot 5 \times 10^{16}$; $\xi_{uv}(water) \approx 1 \cdot 9 \times 10^{16}$.) In this region $\varepsilon_3 \approx 2 > \varepsilon_1 \approx (1 \cdot 333)^2$, and the contribution to μ is positive.

(3) $\xi > \xi_{uv}(hc)$. This region yields another negative contribution to μ since the hydrocarbon has essentially relaxed before water. This is illustrated in Fig. 3.7 where the dielectric function for water is compared with that for various hydrocarbons with a particular interpolation. Clearly the sign of the total chemical potential is not at all obvious.

A rough idea of the total magnitude of the chemical potential can be obtained as follows. Suppose we ignore microwave contributions to the water susceptibility, and model water with two absorption frequencies, ω_1 in the infrared, and ω_2 in the ultraviolet, with oscillator strengths f_1 and f_2. Thus, we write

$$\varepsilon_1 = 1 + \alpha_1(i\xi); \quad \alpha_1(i\xi) = \frac{f_1}{1 + \xi^2/\omega_1^2} + \frac{f_2}{1 + \xi^2/\omega_2^2}. \tag{3.53}$$

The hydrocarbon has one absorption in the ultraviolet, so $\xi_3 = 1 + \alpha_3$

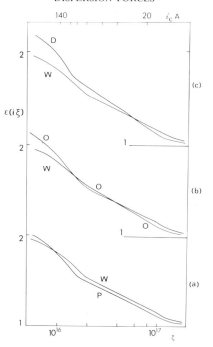

FIG. 3.7. The dielectric susceptibilities of pentane (a), octane (b) and dodecane (c) compared with that of water. The interpolation for water in the uv is the same for each case ranging from the first ionization frequency to 10^{17} rad/sec $[\omega_p(\text{water}) \approx 3 \times 10^{16}$ rad/sec]. Similarly for the hydrocarbons. Top abscissa shows the film thickness l_s for which frequencies $\xi = \xi_s = c/2l\sqrt{\varepsilon_3}$ are damped by retardation. [From Richmond, Ninham and Ottewill (1973).]

where

$$\alpha_3 = \frac{f_3}{1 + \xi^2/\omega_3^2}. \qquad (3.54)$$

We ignore far ultraviolet contributions, which are in any case damped by retardation. Then from (3.52)

$$\mu_{\text{vw}}(l, 0) \underset{l \to 0}{=} \frac{\hbar}{8\pi^2 l^3} \int_0^\infty \left(\frac{\varepsilon_1 - \varepsilon_3}{\varepsilon_1 + \varepsilon_3}\right)\left(\frac{\varepsilon_2 - \varepsilon_3}{\varepsilon_2 + \varepsilon_3}\right) d\xi. \qquad (3.55)$$

This can be reduced to the Hamaker form by replacing the denominators by $\varepsilon_1 + \varepsilon_3 = \varepsilon_2 + \varepsilon_3 \approx 2$, and from (3.53), (3.54) we have

$$\mu_{\text{vw}}(l, 0) = \frac{\hbar}{32\pi^2 l^2} \int_0^\infty d\xi(\alpha_1\alpha_3 - \alpha_3^2)$$

$$= \frac{\hbar\omega_3}{64\pi l^3} f_3^2\left(1 - \frac{f_2}{f_3}\frac{2\omega_2}{\omega_2 + \omega_3} - \frac{f_1}{f_2}\frac{2\omega_1}{\omega_1 + \omega_3}\right). \qquad (3.56)$$

The first two terms correspond to those which would be obtained on the basis of London interactions in the ultraviolet. The third term arises from correlations between molecular vibrations in water and the atomic (electronic) vibrations in hydrocarbon. Usually this latter term is unimportant, but for this system the first and second terms almost cancel. The oscillator strengths can be obtained from refractive index data. For water we have $f_1 \approx 5 \cdot 2 - n^2 = 3 \cdot 42$, $f_2 = (1 \cdot 33)^2 - 1 = 0 \cdot 78$. For pentane, from Table 3.2, we have $f_3 = (1.36)^2 - 1 = 0 \cdot 85$. Evaluating (3.56) we have

$$\mu_{vw}(l, 0) \approx \left\{ \frac{\hbar \omega_3 f_3^3}{64 \pi l^3} \right\} \{1 \cdot 0 - 1 \cdot 0 - 0 \cdot 28\}.$$

Notice that the first two terms cancel, and the stability of a pentane film is determined by the contribution from the third term arising from infrared frequencies. For the higher hydrocarbons the second term of (3.56) decreases while the third term remains essentially unchanged. Thus we can expect the chemical potential will change sign. This crude analysis indicates that the sign change takes place at dodecane, but the approximate results are altogether too crude. Calculations from the full expression of eqn. (3.51) have been carried out by Richmond, Ninham and Ottewill (1973) (see also Kruglyakov, 1974) and indicate that the sign change takes place at octane in agreement with observation (Fig. 3.8). The

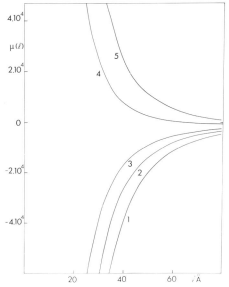

FIG. 3.8. Results for the chemical potential of pentane (1), hexane (2), heptane (3), octane (4) and dodecane (5) films on pure water. [From Richmond, Ninham and Ottewill (1973).]

calculations on dodecane show some interesting features not exhibited by the lower hydrocarbons. The chemical potential is here positive (so that no film forms) due to a large positive contribution from the optical region. However at larger distances, $l > 100$ Å, these frequencies are strongly damped by retardation, and at $l \sim 150$ Å the chemical potential changes sign. A curve of this nature can in principle give rise to wetting at some large thickness l. This and other more interesting possibilities have been discussed by Dzyaloshinskii, Lifshitz and Pitaevskii (1961). A detailed study of the spreading of hydrocarbon films (200–1000 Å) on smooth saphire surfaces has been carried out by Blake (1975) and Richmond (1975).

The inverse problem, the spreading of water on surfaces is a very different and much more difficult matter. Here van der Waals forces do not play the sole determining role. The process depends on the degree of hydrophobicity (due to an adsorbed monolayer or few layers of water) on the surface (Laskowski and Kitchener, 1969).

A further prediction of the theory, not yet confirmed, is that the addition of salt will act to stabilize films. The main effect of salt is to increase the refractive index of water, thereby reducing the positive optical contributions. Calculations show that if 2M NaCl is dissolved in water octane will form a stable film, while 5M salt stabilizes dodecane.

Work carried out by Parsegian and Gingell (1973) (see also Gingell and Parsegian, 1973) on the adhesion of model biological cell surfaces to various substrates reveal the same behaviour—as two macroscopic bodies come together, depending on material properties and geometry, the van der Waals forces can be repulsive at large distances, becoming attractive at smaller distances and vice versa. This remarkably diverse behaviour of van der Waals forces as predicted by the Lifshitz theory, due to the complex interplay of material properties, geometry, temperature-dependence, infrared, optical and ultraviolet fluctuations, and of retardation, must play some part in developing notions concerning the specificity of biological interactions.

3.6 Force between an atom and a macroscopic body

Very little work has been done on direct measurements of the force between an atom and a macroscopic body. Some recent experiments on deflection of C_s atomic beams by surfaces (gold-coated glass, glass and stainless steel) give definite confirmation of the $(1/R^3)$ law for the interaction potential between the atom and the surface (Raskin and Kusch, 1969; Shih, 1974). These experimental results have been analysed in terms of the image-interaction theories of Mavroyannis (1963) and

Bardeen (1940). It is possible that the results may agree better with a microscopic theory that goes into a more detailed analysis of the dispersion interaction between an atom and a surface.

References

Abrikosov, A. A., Gor'kov, L. P. and Dzyaloshinskii, I. E. (1965). "Quantum Field Theoretical Methods in Statistical Physics," 2nd ed. Pergamon Press, London.

Anderson, O. T., Liebenberg, D. H. and Dillinger, J. R. (1960). *Phys. Rev.* **117,** 39.

Anderson, C. H. and Sabisky, E. S. (1970). *Phys. Rev. Letts.* **24,** 1049.

Bailey, A. I. and Daniels, H. (1973). *J. Phys. Chem.* **77,** 501.

Bardeen, J. (1940). *Phys. Rev.* **58,** 727.

Bargeman, J. (1972). *J. Coll. Interface Sci.* **40,** 344.

Black, W., de Jongh, J. V. G., Overbeek, J. Th. G. and Sparnaay, M. J. (1960). *Trans. Farad. Soc.* **56,** 1597.

Blake, T. D. (1975). *J. Chem. Soc. Farad. I* **71,** 192.

Bruil, H. G. (1970). "Specific Ionic Effects in Free Liquid Films," Ph.D. Thesis, Agricultural University, Wageningen, Netherlands.

Deryaguin, B. V. and Abrikosova, I. I. (1956). *J. Exp. Theor. Phys.* **30,** 993; **31,** 3 (*Sov. Phys. JETP* **3,** 819; **4,** 2 (1957)).
  See also Deryaguin, B. V., Abrikosova, I. I. and Lifshitz, E. M. (1956). *Quart. Rev. Chem. Soc.* **10,** 295.

Dzyaloshinskii, I. E., Lifshitz, E. M. and Pitaevskii, L. P. (1961). *Advan. Phys.* **10,** 165.

Fowkes, F. M. (1968). *J. Coll. Interface Sci.* **28,** 493.

Fröhlich, H. (1958). "Theory of Dielectrics," Oxford U.P., New York.

Gingell, D. and Parsegian, V. A. (1972). *J. theoret. Biol.* **36,** 41.

Gingell, D. and Parsegian, V. A. (1973). *J. Coll. Interface Sci.* **44,** 456.

Gregory, J. (1969). *Adv. Coll. Interface Sci.* **2,** 396.

Hauxwell, F. and Ottewill, R. H. (1970). *J. Coll. Interface Sci.* **34,** 473.

Haydon, D. A. and Taylor, J. L. (1968). *Nature* **217,** 739.

Hunklinger, S., Geisselmann, H. and Arnold, W. (1972). *Rev. Sci. Instruments* **43,** 584.

Israelachvili, J. N. (1973). *J. Chem. Soc., Farad. II* **69,** 1729.

Israelachvili, J. N. and Tabor, D. (1972). *Proc. Roy. Soc. A* **331,** 19.

Israelachvili, J. N. and Tabor, D. (1973). *Progress in Surface and Membrane Science* **7,** 1.

Kirkwood, J. G. and Shumaker, J. (1952). *Proc. Nat. Acad. Sci. USA* **38,** 863.

Kitchener, J. A. and Prosser, A. D. (1957). *Proc. Roy. Soc. A* **242,** 403.

Kruglyakov, P. M. (1974). *Kolloid. Zhurnal* **36,** 160.

Krupp, H. (1967). *Advan. Coll. Interface Sci.* **1,** 111.

Krupp, H., Schnabel, W. and Walter, J. (1972). *J. Coll. Interface Sci.* **39,** 421.

Landau, L. D. and Lifshitz, E. M. (1958). "Statistical Physics." Pergamon, London.

Landau, L. D. and Lifshitz, E. M. (1960). "Electrodynamics of Continuous Media." Addison–Wesley Publ. Co., Reading, Mass.

Langbein, D. (1969). *J. Adhesion* **1,** 237.

Langbein, D. (1971). *J. Phys. Chem. Solids* **32,** 1657.

Laskowski, J. and Kitchener, J. A. (1969). *J. Coll. Interface Sci.* **29,** 670.

Lyklema, J. and Mysels, K. J. (1965). *J. Am. Chem. Soc.* **87,** 2539.

Mavroyannis, C. (1963). *Mol. Phys.* **6,** 593.

Ninham, B. W. and Parsegian, V. A. (1970a). *J. Chem. Phys.* **52,** 4578.
  See also Parsegian, V. A. and Ninham, B. W. (1969). *Nature (London)* **227,** 739.

Ninham, B. W. and Parsegian, V. A. (1970b). *Biophys. J.* **10,** 646.

Nir, S., Rein, R. and Weiss, L. (1972). *J. theoret. Biol.* **34,** 135.

Ottewill, R. H. and Shaw, J. N. (1966). *Disc. Farad. Soc.* **42,** 154.

Parsegian, V. A. (1970). "NAS-NRC 1968 Digest of Literature on Dielectrics," R. E. Barker Jr. (ed.). Nat. Acad. of Sciences, Washington DC. **32,** 285.

Parsegian, V. A. and Ninham, B. W. (1970). *Biophys. J.* **10,** 664.

Parsegian, V. A. and Gingell, D. (1973). "A Physical Force Model of Biological Membrane Interaction" in *Recent Advances in Adhesion* (ed. H. L. Lee), Gordon and Breach, London. pp. 153–192.

Pitaevskii, L. P. (1959). *J. Exp. Theoret. Phys.* **37,** 577 (*Sov. Phys. JETP* **10,** 408).

Raskin, D. and Kusch, P. (1969). *Phys. Rev.* **179,** 712.

Richmond, P. (1975). "The Theory and Calculation of van der Waals Forces" in *Specialist periodical reports "Colloid Science"*, vol. 2 (Ed. D. H. Everett), Chemical Society.

Richmond, P. and Ninham, B. W. (1971). *J. Low Temp. Phys.* **5,** 177.

Richmond, P. and Ninham, B. W. (1972). *J. Coll. Interface Sci.* **40,** 406.

Richmond, P., Ninham, B. W. and Ottewill, R. H. (1973). *J. Coll. Interface Sci.* **45,** 69.

Rouweler, G. C. J. and Overbeek, J. Th. G. (1971). *Trans. Farad. Soc.* **67,** 2117.

Sabisky, E. S. and Anderson, C. H. (1973). *Phys. Rev. A* **7,** 790.

Shih, A., Raskin, D. and Kusch, P. (1974). *Phys. Rev. A* **9,** 652.

Tabor, D. and Winterton, R. H. S. (1969). *Proc. Roy. Soc. A* **312,** 435.

van den Tempel, M. (1972). *Advan. Coll. Interface Sci.* **3,** 137.

van Silfhout, A. (1966). *Kon. Ned. Akad. v. Wetensk. Proc. B* **69,** 501, 517, 533.

van Voorst Vader, F. (1972). Proc. 6th International Congress on Surface Active Substances, Zürich (Sept.).

Watillon, A. and Joseph–Petit, A. M. (1966). *Disc. Farad. Soc.* **42,** 143.

Wittmann, F., Splittgerber, H. and Ebert, K. (1971). *Z. Physik* **245,** 354.

Data sources for dielectrics

The Handbook of Chemistry and Physics, late editions. Chemical Rubber Publishing Co., Cleveland, Ohio.

Tables of Dielectric Data for Pure Liquids and Dilute Solutions. Natl. Bur. Stds. (U.S.A.), circular **589,** 7 (1958).

Selected Values of the Physical Properties of Hydrocarbons and Related Compounds (ed. Rossini, Pitzer and Arnett). American Petroleum Institute Program, Thermodynamics Research Center, College Station, Texas (1967).

Hirayama, K. (1967). "Handbook of Ultraviolet and Visible Absorption Spectra of Organic Compounds." Plenum Press Data Division, New York.

Kislovskii, L. D. (1956). *Optics and Spectroscopy* **1,** 672.

Kislovskii, L. D. (1957). *Optics and Spectroscopy* **2,** 186.

Kislovskii, L. D. (1959). *Optics and Spectroscopy* **7,** 201.

Chapter 4

Dispersion Forces and Molecular Size

4.1 Characterization of molecular size

The theory of dispersion forces developed so far is based on the concept of the interaction of point molecules with the radiation field. This introduces divergences in the theory. Examples are the divergence in the dispersion self-energy of a molecule, and that in the Lifshitz interaction energy between two dielectric slabs at zero separation, which has been mentioned earlier. In practice of course, the size of the molecules, although small, will have significant effects at separations comparable with it. In this chapter we shall investigate various aspects of the effect of molecular size on dispersion forces.

London (1942) had developed an extension of his theory of dispersion forces between point molecules, to consider the forces between anisotropic force centres, and between molecules containing extended electronic oscillators (charge transfer spectra). During more recent years the dispersion forces between large molecules have received considerable attention and the work in this field has been reviewed by Linder and Rabenold (1972), who have given a unified treatment of van der Waals forces between two molecules of arbitrary sizes and electron delocalizations. Their approach based on linear response theory synthesizes the earlier approaches which are based on formulation of the interaction (i) in terms of charge fluctuations on individual molecules (McLachlan, 1963a, b; Linder, 1964; Jehle, 1965), (ii) in terms of the collective properties of the interacting bodies (Lundqvist and Sjölander, 1964; Mahan, 1965; Lucas, 1967), and finally, (iii) in terms of the collective properties of different parts of the interacting bodies, such as is implied in the theory of Lifshitz or that of Langbein reported in Chapter 2. We shall formulate the problem here in a manner that logically extends the field approach developed in Chapter 2.

Let us consider a single atom interacting with the radiation field. Qualitatively, the effect of the size of the atom appears when we consider the polarization induced on it due to an electric field in the form of a plane wave. If the wave number is larger than the reciprocal of the linear size of the atom, different parts of the atom will experience different fields (in fact, opposing fields), and the net polarization developed will be small. We may expect it to tend to zero as the wave number $k \to \infty$. For a point atom, on the other hand, no such effect need be expected. Quantitatively, this can be expressed in terms of a k-dependent polarizability tensor $\boldsymbol{\alpha}_n(\mathbf{k}, \omega)$, which gives the polarization of the atom in its n-th state due to an incident electric field of the form

$$\mathbf{E}(\mathbf{r}, t) = \mathbf{E}(\mathbf{k}, \omega)e^{i(\mathbf{k}\cdot\mathbf{r} - \omega t)}, \qquad (4.1)$$

the amplitude of the polarization being,

$$\mathbf{P}_n(\mathbf{k}, \omega) = [\boldsymbol{\alpha}_n(\mathbf{k}, \omega)\mathbf{E}(\mathbf{k}, \omega)], \qquad (4.2)$$

The evaluation of $\boldsymbol{\alpha}_n(\mathbf{k}, \omega)$ can be done by use of linear response theory as in Chapter 2. For a single-electron atom, for instance, the polarization operator in terms of the electron coordinate $\mathbf{u}$ measured from the nucleus (which is at $\mathbf{R}$), is given by

$$\mathbf{p} \equiv -e\mathbf{u}. \qquad (4.3)$$

The perturbation Hamiltonian due to the electric field of (4.1) is approximately given by taking the electric field at the middle of the dipole, and multiplying with the dipole moment,

$$H_I(t) \equiv [e\mathbf{u}\cdot\mathbf{E}(\mathbf{k}, \omega)e^{i\mathbf{k}\cdot\mathbf{u}/2}]e^{i(\mathbf{k}\cdot\mathbf{R} - \omega t)}. \qquad (4.4)$$

A more exact treatment will be given in Appendix B. The polarization developed on the atom in the state $|n\rangle$ due to this perturbation is obtained by using linear response theory (Kubo, 1957), in the form,

$$\mathbf{P}_n(t) = \frac{1}{i\hbar} \int_{-\infty}^{t} \langle n|[\mathbf{p}(t - t'), H_I(t')]|n\rangle \, dt', \qquad (4.5)$$

where $\mathbf{p}(t - t')$ is the polarization operator in the Heisenberg representation,

$$\mathbf{p}(t) = e^{i(H_0 t/\hbar)}(\mathbf{p})e^{-i(H_0 t/\hbar)}, \qquad (4.6)$$

H_0 being the unperturbed atomic Hamiltonian. Using (4.4) and (4.3) in (4.5), doing the time integration and after some simplifications we get (Mahanty and Ninham, 1973; Mahanty, 1974),

$$\mathbf{P}_n(t) = [\boldsymbol{\alpha}_n(\mathbf{k}, \omega)\mathbf{E}(\mathbf{k}, \omega)]e^{i(\mathbf{k}\cdot\mathbf{r} - \omega t)}, \tag{4.7}$$

where the polarizability tensor $\boldsymbol{\alpha}_n(\mathbf{k}, \omega)$ in the n-th state of the atom is given by

$$\boldsymbol{\alpha}_n(\mathbf{k}, \omega) = -\frac{e^2}{\hbar}\sum_m \left(\frac{\{\langle n|\,\mathbf{u}\,|m\rangle\langle m|\,\mathbf{u}e^{i\mathbf{k}\cdot\mathbf{u}/2}|n\rangle\}}{(\omega_{nm} + \omega)} + \frac{\{\langle m|\,\mathbf{u}\,|n\rangle\langle n|\,\mathbf{u}e^{i\mathbf{k}\cdot\mathbf{u}/2}|m\rangle\}}{(\omega_{nm} - \omega)}\right). \tag{4.8}$$

Here $\{\langle n|\,\mathbf{u}\,|m\rangle\langle m|\,\mathbf{u}e^{i\mathbf{k}\cdot\mathbf{u}/2}|n\rangle\}$ stands for the dyadic formed out of the two vector matrix elements and

$$\omega_{nm} = \frac{E_n - E_m}{\hbar}.$$

Equation (4.7) immediately leads to the result that the polarization due to an arbitrary field $\mathbf{E}(\mathbf{r}, \omega)$, which can always be expressed as a linear combination of $\mathbf{E}(\mathbf{k}, \omega)e^{i(\mathbf{k}\cdot\mathbf{r})}$, will be given by

$$\mathbf{P}_n(\omega) = \int d^3k\,\boldsymbol{\alpha}_n(\mathbf{k}, \omega)\mathbf{E}(\mathbf{k}, \omega)e^{i\mathbf{k}\cdot\mathbf{R}}$$

$$= \int d^3r\,\boldsymbol{\alpha}_n(\mathbf{r} - \mathbf{R}; \omega)\mathbf{E}(\mathbf{r}, \omega), \tag{4.9}$$

where $\boldsymbol{\alpha}_n(\mathbf{r}, \omega)$ is the Fourier transform of $\boldsymbol{\alpha}(\mathbf{k}, \omega)$. The quantity $\boldsymbol{\alpha}_n(\mathbf{r} - \mathbf{R}; \omega)\mathbf{E}(\mathbf{r}, \omega)$ can thus be interpreted as a polarization density $\mathbf{p}_n(\mathbf{r}, \omega)$ whose integral over all space gives the total polarization developed by the atom.

From the explicit form of $\boldsymbol{\alpha}(\mathbf{k}, \omega)$ in (4.8) we note that its $\mathbf{k}$ dependence arises through the factor

$$\langle m|\,\mathbf{u}e^{i\mathbf{k}\cdot\mathbf{u}/2}\,|n\rangle \equiv -(i\nabla_k)\int d^3u\,\Phi_m(\mathbf{u})\Phi_n(\mathbf{u})e^{i\mathbf{k}\cdot\mathbf{u}/2}. \tag{4.10}$$

For $k \rightarrow \infty$ this must tend to zero. Since the atomic wave function $\Phi_n(\mathbf{u})$ is peaked with a finite range, one would expect the above matrix element to be peaked around $k = 0$ with a range that is of the order of the reciprocal of the range of $\Phi_n(\mathbf{u})$, i.e., the size of the atom in the n-th state. This property must persist in $\boldsymbol{\alpha}_n(\mathbf{k}, \omega)$, whence $\boldsymbol{\alpha}_n(\mathbf{r} - \mathbf{R}; \omega)$ will be a peaked function with the peak around $\mathbf{R}$ and a range of the order of the atomic size. Because of the peaked nature of $\boldsymbol{\alpha}_n(\mathbf{r} - \mathbf{R}; \omega)$, we can also write

$$\mathbf{p}_n(\mathbf{r}, \omega) \equiv \boldsymbol{\alpha}_n(\mathbf{r} - \mathbf{R}; \omega)\mathbf{E}(\mathbf{r}, \omega) \cong \boldsymbol{\alpha}_n(\mathbf{r} - \mathbf{R}; \omega)\mathbf{E}(\mathbf{R}, \omega). \tag{4.11}$$

For a point atom, the term in $[e^{i\mathbf{k}.\mathbf{u}/2}]$ does not occur in the matrix element, so that $\boldsymbol{\alpha}_n(\mathbf{k}, \omega)$ is independent of $\mathbf{k}$. This leads to $\boldsymbol{\alpha}_n(\mathbf{r}-\mathbf{R}; \omega)$ becoming a δ-function $\boldsymbol{\alpha}(\omega)\delta(\mathbf{r}-\mathbf{R})$.

We thus arrive at the conclusion that the size of an atomic system, i.e., an atom or a molecule can be characterized in terms of a polarizability tensor $\boldsymbol{\alpha}_n(\mathbf{r}-\mathbf{R}; \omega)$ which leads to a polarization density according to (4.11) having a range of the order of the size of the system.

The same kind of reasoning can be used to establish that there will be a size distribution associated with each of the other multipole moments induced on an atomic system by an external field (Richardson, 1975). We shall consider here the dipole polarizability only, that being the dominant one in dispersion interactions.

4.2   The dispersion self-energy of the atomic system and interaction energies

In this section we shall obtain an expression for the dispersion self-energy of the atomic system, which is the change in the zero-point energy of the electromagnetic field due to its coupling with the atomic system. We start with the field equations in time-independent (Fourier transform) form. In Lorentz gauge, the equation for the vector potential can be written as

$$\left(\nabla^2 + \frac{\omega^2}{c^2}\right)\mathbf{A}(\mathbf{r}, \omega) = -4\pi i\left(\frac{\omega}{c}\right)\mathbf{p}(\mathbf{r}, \omega), \qquad (4.12)$$

and the electric field is given by

$$\mathbf{E}(\mathbf{r}, \omega) = -\frac{ic}{\omega}\left(\frac{\omega^2}{c^2}\mathbf{A} + \nabla(\nabla.\mathbf{A})\right). \qquad (4.13)$$

The relationship between $\mathbf{E}(\mathbf{r}, \omega)$ and $\mathbf{p}(\mathbf{r}, \omega)$ has already been indicated in the previous section. Identifying $\mathbf{p}(\mathbf{r}, \omega)$ in (4.12) with $\mathbf{p}_n$ of (4.11) we can write (4.12) in the form of an integral equation

$$\mathbf{A}(\mathbf{r}, \omega) = -4\pi\left(\int G(\mathbf{r}-\mathbf{r}'; \omega)\boldsymbol{\alpha}(\mathbf{r}'-\mathbf{R}; \omega)\,\mathrm{d}^3r'\right)$$
$$\times\left(\frac{\omega^2}{c^2}\mathbf{A}(\mathbf{R}; \omega) + \nabla(\nabla.\mathbf{A})_{\mathbf{r}=\mathbf{R}}\right), \quad (4.14)$$

where the dyadic Green function $G(\mathbf{r}-\mathbf{r}'; \omega)$ is,

$$G(\mathbf{r}-\mathbf{r}'; \omega) = \mathscr{I}\frac{1}{(2\pi)^3}\int\frac{e^{i\mathbf{k}.(\mathbf{r}-\mathbf{r}')}}{(\omega^2/c^2) - k^3}\,\mathrm{d}^3k. \qquad (4.15)$$

From (4.13) and (4.14) we get the secular equation for the perturbed frequencies of the field,

$$|\mathscr{I} + 4\pi\mathscr{G}(\mathbf{R}, \mathbf{R}; \omega)| = 0, \tag{4.16}$$

where

$$\mathscr{G}(\mathbf{r}, \mathbf{r}'; \omega) = \left(\frac{\omega^2}{c^2}\mathscr{I} + (\nabla_r\nabla_r)\right)\int G(\mathbf{r} - \mathbf{r}''; \omega)\boldsymbol{\alpha}(\mathbf{r}'' - \mathbf{r}'; \omega)\, \mathrm{d}^3 r''$$

$$\equiv \int \mathscr{G}(\mathbf{r}, \mathbf{r}''; \omega)\boldsymbol{\alpha}(\mathbf{r}'' - \mathbf{r}'; \omega)\, \mathrm{d}^3 r''. \tag{4.17}$$

Here $(\nabla_r\nabla_r)$ is the dyadic operator formed out of the gradient.

As in Chapter 2, we can write the dispersion self-energy of the atomic system in the form of a contour integral,

$$E_s = \frac{1}{2\pi i}\oint\left(\frac{\hbar\omega}{2}\right)\frac{\mathrm{d}}{\mathrm{d}\omega}\ln|\mathscr{I} + 4\pi\mathscr{G}(\mathbf{R}, \mathbf{R}; \omega)|\, \mathrm{d}\omega$$

$$= -\frac{\hbar}{4\pi i}\oint \mathrm{d}\omega \ln|\mathscr{I} + 4\pi\mathscr{G}(\mathbf{R}, \mathbf{R}; \omega)|$$

$$= \frac{\hbar}{4\pi}\int_{-\infty}^{\infty} \mathrm{d}\xi \ln|\mathscr{I} + 4\pi\mathscr{G}(\mathbf{R}, \mathbf{R}; i\xi)|. \tag{4.18}$$

The transformation of the contour integral (4.18) to the last form has already been discussed in §1.7 in connection with eqn. (1.75).

It may be noticed from the definition of $\mathscr{G}$ in (4.17) that the elements of

$$\mathscr{G}(\mathbf{R}, \mathbf{R}; \omega) = \lim_{r \to R} \mathscr{G}(\mathbf{r}, \mathbf{R}; \omega)$$

are not likely to diverge, although those of $\mathscr{G}(\mathbf{R}, \mathbf{R}; \omega)$ obtained from (4.15) and (4.17) do diverge. The latter singularity is responsible for the infinite dispersion self-energies of point molecules. The finite size of the atomic system occurring in the function $\boldsymbol{\alpha}(\mathbf{r} - \mathbf{R}; \omega)$ is responsible for a finite value of its self-energy. To leading order in the polarizability we get

$$E_s = \hbar\int_{-\infty}^{\infty} \mathrm{d}\xi \operatorname{Tr} \mathscr{G}(\mathbf{R}, \mathbf{R}; i\xi). \tag{4.19}$$

Further computation requires knowledge of the explicit forms of $\boldsymbol{\alpha}(\mathbf{r}, \omega)$, which in principle can be evaluated from the wave functions of the atomic system as discussed in the previous section. To bring out the physical concepts involved, we will assume that the atomic system is isotropic and that $\boldsymbol{\alpha}(\mathbf{r}, \omega)$ has the form

$$\boldsymbol{\alpha}(\mathbf{r}, \omega) \equiv \mathscr{I}\alpha(\omega)f(r), \tag{4.20}$$

where $f(r)$ is a peaked function. We shall further assume that f is a Gaussian function with width a,

$$f(r) = \frac{1}{\pi^{3/2} a^3} e^{-r^2/a^2} \tag{4.21}$$

where a is the size of the atomic system. With this choice we obtain (Mahanty and Ninham, 1973; Mahanty, 1974)

$$\begin{aligned}
\operatorname{Tr} \mathscr{G}(\mathbf{R}, \mathbf{R}; i\xi) &= \frac{\alpha(i\xi)}{(2\pi)^3} \int \frac{d^3 k [3\xi^2/c^2 + k^2] \exp(-k^2 a^2/4)}{[\xi^2/c^2 + k^2]} \\
&= \frac{\alpha(i\xi)}{\pi^{3/2} a^3} + \frac{4\alpha(i\xi)}{\pi a^3} \left\{ \frac{1}{\sqrt{\pi}} \left(\frac{a\xi}{2c}\right)^2 \right. \\
&\quad \left. - \left(\frac{a\xi}{2c}\right)^2 e^{-(a\xi/2c)^2} \operatorname{erfc}\left(\frac{a\xi}{2c}\right) \right\}. \tag{4.22}
\end{aligned}$$

Here erfc is the complementary error function, defined by

$$\operatorname{erfc}(z) = \frac{2}{\sqrt{\pi}} \int_z^\infty e^{-t^2} \, dt.$$

When this expression is substituted into (4.19), the main contribution to the ξ-integration will come from the characteristic absorption frequencies of the atomic system that determine the poles of $\alpha(\omega)$. For the corresponding value of ξ, $(a\xi/2c) \cong (a/2\lambda_{ch})$, where λ_{ch} is the wavelength of a characteristic absorption line. This ratio will generally be negligible, since a would be of the order of a few Ångströms at most, while λ_{ch} is much larger. Hence, for all practical purposes, for determination of the *self-energy* of an atomic system we can use the approximate formulae,

$$\operatorname{Tr} \mathscr{G}(\mathbf{R}, \mathbf{R}; i\xi) \cong \frac{\alpha(i\xi)}{\pi^{3/2} a^3} \tag{4.23}$$

and

$$E_s \cong \frac{\hbar}{\pi^{3/2} a^3} \int_{-\infty}^\infty d\xi \, \alpha(i\xi). \tag{4.24}$$

This rather over-simplified example serves to bring out the main point that the size of the atomic system enters the expression for its dispersion self-energy in a very direct way. The dispersion self-energy can be expected to depend on the inverse cube of the size of the atomic system. As expected, in free space, E_s does not depend on $\mathbf{R}$, the location of the atomic system. A point to be noticed is that (4.23) arises only from the non-retarded form $(c \to \infty)$ of $\mathscr{G}(\mathbf{r} - \mathbf{r}'; \omega)$, so that for the purpose of estimating the self-energy the non-retarded Green function only need be used.

For the simple classical form $\alpha(i\xi) = e^2/m(\omega_0^2 + \xi^2)$, eqn. (4.24) reduces to $E_s \approx \hbar e^2/\sqrt{\pi} a^2 m\omega_0$. If we make the identification $\hbar\omega_0 \approx |E_G| = e^2/2a_0$, where E_G is the ground state energy for hydrogen, and take $a = a_0$, the Bohr radius for a hydrogen atom, we see that $E_s = (2/\sqrt{\pi})(e^2/a_0) = 4/\sqrt{\pi}$ (Ry) i.e., the self-energy of an atom is of the same order of magnitude as its binding energy, but of the opposite sign.

It is easy to extend the formalism developed here to consider the interaction energy of two atomic systems. The procedure is exactly as was followed in getting eqn. (2.33). If we consider two atomic systems with polarizabilities $\boldsymbol{\alpha}_1(\mathbf{r} - \mathbf{R}_1; \omega)$ and $\boldsymbol{\alpha}_2(\mathbf{r} - \mathbf{R}_2; \omega)$, the analogue of (2.28) would be,

$$\mathbf{E}(\mathbf{r}, \omega) = -4\pi\left\{\left(\int \mathscr{G}(\mathbf{r} - \mathbf{r}'; \omega)\boldsymbol{\alpha}_1(\mathbf{r}' - \mathbf{R}_1; \omega)\, d^3r'\right)\right.$$

$$\times \mathbf{E}(\mathbf{R}_1, \omega) + \left.\left(\int \mathscr{G}(\mathbf{r} - \mathbf{r}'; \omega)\boldsymbol{\alpha}_2(\mathbf{r}' - \mathbf{R}_2; \omega)\, d^3r'\right)\mathbf{E}(\mathbf{R}_2, \omega)\right\}$$

$$= -4\pi[\mathscr{G}_1(\mathbf{r}, \mathbf{R}_1; \omega)\mathbf{E}(\mathbf{R}_1; \omega) + \mathscr{G}_2(\mathbf{r}, \mathbf{R}_2; \omega)\mathbf{E}(\mathbf{R}_2; \omega)], \quad (4.25)$$

where the subscripts 1 and 2 in the Green functions $\mathscr{G}_1$ and $\mathscr{G}_2$ as defined in (4.25) represent different polarizability tensors corresponding to the two systems. The secular determinant analogue of eqn. (2.33) is now easily constructed, and the interaction energy is now given by

$$V(R_{12}) \cong -4\pi\hbar \int_{-\infty}^{\infty} d\xi\, \mathrm{Tr}\,\{\mathscr{G}_2(\mathbf{R}_1, \mathbf{R}_2; i\xi)\mathscr{G}_1(\mathbf{R}_2, \mathbf{R}_1; i\xi)\}. \quad (4.26)$$

The explicit form of the trace is obtained as follows:

$$\mathscr{G}_1(\mathbf{r}, \mathbf{r}'; i\xi) = \frac{1}{(2\pi)^3}\int \frac{d^3k[(\xi^2/c^2)\mathscr{I} + \mathbf{kk}]}{[(\xi^2/c^2) + k^2]}$$

$$\times e^{+i\mathbf{k}\cdot\mathbf{r}}\int e^{-i\mathbf{k}\cdot\mathbf{r}''}\boldsymbol{\alpha}_1(\mathbf{r}'' - \mathbf{r}', \omega)\, d^3r''$$

$$= \frac{1}{(2\pi)^3}\int \frac{d^3k[(\xi^2/c^2)\mathscr{I} + \mathbf{kk}]}{[(\xi^2/c^2) + k^2]}$$

$$\times e^{i\mathbf{k}\cdot(\mathbf{r} - \mathbf{r}')}\int e^{-i\mathbf{k}\cdot\mathbf{r}'''}\boldsymbol{\alpha}_1(\mathbf{r}''')\, d^3r'''. \quad (4.27)$$

The last factor in the integrand, which is $\boldsymbol{\alpha}_1(\mathbf{k}, \omega)$, is peaked with a range $k_c \sim 1/a_1$. If $|\mathbf{r} - \mathbf{r}'| \gg a_1$, the main contribution to the integral will come from the neighbourhood of $k \to 0$, and hence we get a good approximation by setting $\boldsymbol{\alpha}_1(\mathbf{k}, \omega) \equiv \boldsymbol{\alpha}_1(0; \omega) \equiv \boldsymbol{\alpha}_1(\omega)$, where the latter is the polarizability tensor of the atomic system regarded as a point, as in eqn. (2.37). Similar analysis can be made of $\mathscr{G}_2$, and we thus establish that for

$|\mathbf{R}_1 - \mathbf{R}_2| \gg a_1, a_2$, the sizes of the atomic systems, the results that follow from (2.37) are valid. This is, of course, always the case in the non-retarded and retarded regions with actual atomic systems. However, when $|\mathbf{R}_1 - \mathbf{R}_2|$ becomes comparable with the sizes, (2.37) leads to a strong divergence, whereas (4.26) remains finite, even up to $\mathbf{R}_1 = \mathbf{R}_2$, i.e., even when the two systems overlap. In this region, however, overlap of the electron clouds would render this simple theory untenable. But if the sizes a_1 and a_2, or rather their sum $(a_1 + a_2)$ is treated as a lower bound for $|\mathbf{R}_1 - \mathbf{R}_2|$, we may expect (4.26) to give more reasonable results in this region than (2.37), since the size effects are taken into consideration here. For the "Gaussian molecules" of (4.20) and (4.21), the result is,

$$
\begin{aligned}
V(R_{12}) &\cong -4\pi\hbar \int_{-\infty}^{\infty} d\xi\, \alpha_1(i\xi)\alpha_2(i\xi)\, \mathrm{Tr}\left\{\frac{1}{(2\pi)^6}\int d^3k\, d^3k'\right. \\
&\times \frac{[(\xi^2/c^2)\mathscr{I} + \mathbf{k}\mathbf{k}][(\xi^2/c^2)\mathscr{I} + \mathbf{k}'\mathbf{k}']}{[(\xi^2/c^2) + k^2][(\xi^2/c^2) + k'^2]} \exp(-k^2 a_1^2/4) \cdot \\
&\times \left. \exp(-k'^2 a_2^2/4) e^{i\mathbf{k}\cdot(\mathbf{R}_1-\mathbf{R}_2)} e^{i\mathbf{k}'\cdot(\mathbf{R}_2-\mathbf{R}_1)} \right\} \\
&= -4\pi\hbar \int_{-\infty}^{\infty} d\xi\, \alpha_1(i\xi)\alpha_2(i\xi) \\
&\times \left(2\prod_{j=1}^{2} \frac{\exp(\xi^2 a_j^2/4c^2)}{8\pi R} \left\{ (\xi^2/c^2 + \xi/cR + 1/R^2) F_1^{(j)} \right. \right. \\
&- (\xi^2/c^2 - \xi/cR + 1/R^2) F_2^{(j)} - \frac{4}{a_j R\sqrt{\pi}} \\
&\times \left. \exp(-\xi^2 a_j^2/4c^2 - R^2/a_j^2) \right\} \\
&+ \prod_{j=1}^{2} \frac{\exp(\xi^2 a_j^2/4c^2)}{4\pi R} \left\{ (\xi/cR + 1/R^2) F_1^{(j)} \right. \\
&+ (\xi/cR - 1/R^2) F_2^{(j)} - \frac{4}{a_j\sqrt{\pi}} (1/R + R/a_j^2) \\
&\times \left. \left. \exp(-\xi^2 a_j^2/4c^2 - R^2/a_j^2) \right\} \right),
\end{aligned}
\tag{4.28}
$$

where

$$F_1^{(j)} = \mathrm{erfc}\,(\xi a_j/2c - R/a_j)\exp(-\xi R/c) \tag{4.29a}$$

$$F_2^{(j)} = \mathrm{erfc}\,(\xi a_j/2c + R/a_j)\exp(+\xi R/c), \tag{4.29b}$$

with

$$R \equiv R_{12}.$$

As stated earlier, for $R_{12} \gg a_1, a_2$ this would give the same results as (2.37), i.e., the Casimir–Polder result for the retarded and the London result for the non-retarded case. But in the non-retarded region with $R_{12} \approx a_j$, we get a deviation from the London formula. In the non-retarded region the interaction energy can be written

$$V(R) = V_{\text{London}}(R)F(R), \tag{4.30}$$

where the London potential is

$$V_{\text{London}}(R) = -\frac{3\hbar}{\pi R^6} \int_0^\infty d\xi \, \alpha_1(i\xi)\alpha_2(i\xi), \tag{4.31}$$

$$F(R) = \frac{1}{3} \prod_{j=1}^2 \left\{ \text{erf}\left(\frac{R}{a_j}\right) - \frac{2R}{\sqrt{\pi}a_j} \exp\left(-\frac{R^2}{a_j^2}\right) \right\}$$

$$\times \frac{2}{3} \prod_{j=1}^2 \left\{ \text{erf}\left(\frac{R}{a_j}\right) - \frac{2}{\sqrt{\pi}}\left(\frac{R}{a_j} + \frac{R^3}{a_j^3}\right) \exp\left(-\frac{R^2}{a_j^2}\right) \right\}. \tag{4.32}$$

For $R \gg a_j$, $F(R) \to 1$ as it should, and for $R \to 0$,

$$F(R) \to \frac{8}{9\pi}\left(\frac{R^2}{a_1 a_2}\right)^3,$$

so that

$$\lim_{R \to 0} V(R) = -\frac{8\hbar}{3\pi^2 a_1^3 a_2^3} \int_0^\infty d\xi \, \alpha_1(i\xi)\alpha_2(i\xi). \tag{4.33}$$

This finite limit, although of no great significance, is interesting, since for two like atoms it is of the order of the binding energy of a molecule formed out of the two atomic systems. We note again the important result that the effect of the finite size of the molecules is the removal of the divergence in the London interaction as $R \to 0$.

If we use $V(R)$ of eqn. (4.30) as the form of the basic two-body interaction in an assembly of molecules, the force between two macroscopic bodies can be estimated by the method of pairwise summation. We shall consider the case of two dielectric slabs with their geometry indicated in the inset to Fig. 4.1. Let the densities of the molecules in the two slabs be n_1 and n_2 respectively. Summing the pair interaction given by eqn. (4.30) over all the molecules, and replacing the sums by integrals, we can estimate the interaction energy per unit area as

$$E^A(l) = -\frac{A_{12}}{\pi^2} \int_l^\infty dz \int_0^\infty dz' \int_0^\infty 2\pi\rho \, d\rho \, \frac{F[\sqrt{\rho^2 + (z+z')^2}]}{[\rho^2 + (z+z')^2]^3}, \tag{4.34}$$

where

$$A_{12} = 3\hbar\pi n_1 n_1 \int_0^\infty \alpha_1(i\xi)\alpha_2(i\xi) \, d\xi \tag{4.35}$$

is the Hamaker constant in the notation of Chapter 1. Two of the integrals can be done at once to give

$$E^A(l) = -A_{12}I(l, a_1, a_2),$$ (4.36)

where

$$I(l, a_1, a_2) = \frac{1}{\pi} \int_0^\infty z^2 \, dz \, \frac{F(z+l)}{(z+l)^5}.$$ (4.37)

For "large" separations of the slabs, $l \gg a_j$, we have

$$E^A(l) \approx -\frac{A_{12}}{12\pi l^2}\left\{1 - \frac{4}{\sqrt{\pi}}\left(\left(\frac{a_1}{l}\right)^3 \exp\left(-\frac{l^2}{a_1^2}\right)\right.\right.$$
$$\left.\left. + \left(\frac{a_2}{l}\right)^3 \exp\left(-\frac{l^2}{a_2^2}\right)\right) + \cdots\right\}.$$ (4.38)

Figure 4.1 indicates the nature of the function $I(l)$ computed numerically from eqn. (4.37). The interaction energy is finite at $l = 0$, and some of the asymptotic results for this limit are

$$I(0) = (\pi - 1)/3a_1^2\pi^2 \quad \text{for} \quad a_1 = a_2$$ (4.39)

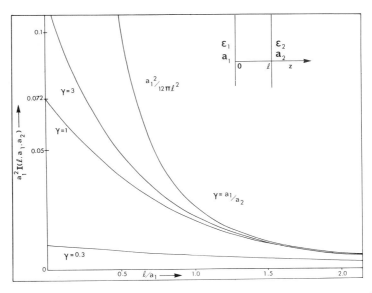

FIG. 4.1. Result of pairwise integration of interaction energies. Interaction energy is $E^A(l) = -A_{12}I(l, a_1, a_2)$ with A_{12} the Hamaker constant. For $l \gg a_1, a_2$, $E^A(l)$ goes over to the Hamaker result $A_{12}/12\pi l^2$. Ordinate plotted is $a_1^2 I(l, a_1, a_2)$ for different size ratios $\gamma = a_1/a_2$. Interaction energy is always finite. [From Mahanty and Ninham (1974).]

and

$$I(0) = \frac{1}{3\pi} \frac{\bar{a}}{a_1 a_2^2} + \frac{4}{9\pi^2} \frac{\bar{a}}{a^3} + \cdots \quad \text{for} \quad a_2 \gg a_1 \qquad (4.40)$$

with $\bar{a}$ defined by $1/\bar{a}^2 = 1/a_1^2 + 1/a_2^2$. It may be noted that even if $a_1 \to 0$, this limit remains finite at $1/3\pi a_2^2$ even though the two-particle interaction of eqn. (4.33) diverges. This is due to the replacement of sums by integrals. But the interesting point that emerges is that the interaction energy of two slabs at zero separation is governed by the one which contains the larger molecules.

These results do not take into account the mutual interactions of the molecules in each slab, and hence must be regarded as an approximation. We remark in passing that although the detailed form of the interaction energy depends on the assumed form of $\boldsymbol{\alpha}(\mathbf{r}; \omega)$, the molecular polarizability density functions, the general nature of the interaction energy and in particular its finiteness at zero separation depends only on the fact that $\boldsymbol{\alpha}(\mathbf{r}; \omega)$ is a peaked function with spread of the order of molecular size. Our choice of the Gaussian form of the distribution is primarily for mathematical convenience.

4.3   A note on radiative corrections in quantum electrodynamics

The concept of dispersion self-energy developed in the previous section can be used to make a semi-classical estimate of radiative corrections to spectral lines (Mahanty, 1974), in a manner that is considerably simpler than calculations of such corrections via quantum electrodynamics. Radiative corrections to the energy levels of an atom and the dispersion interaction between two atoms are two facets of the same physical process, the process being the change in the electromagnetic field modes due to the presence of the atomic systems. We shall give here, for the purpose of illustration, an estimate of the Lamb shift in hydrogen from considerations of the dispersion self-energy of the hydrogen atom.

If eqn. (4.19) is used to evaluate the energy shifts for two different levels, the differences between the shifts would give us the radiative correction to the energy difference. Since the radiative corrections involve differences between two self-energies, regarding the atom as a point particle (for which the dispersion self-energy would diverge in each atomic state) does not introduce difficulties because of cancellation of divergences in taking the difference. We can thus take

$$\boldsymbol{\alpha}_n(\mathbf{k}, \omega) \cong \mathscr{I}\left(\frac{2}{\hbar}\right) \sum_m \frac{|\langle n| \, \mu \, |m\rangle|^2}{\omega^2 - \omega_{nm}^2} \, \omega_{nm}, \qquad (4.41)$$

where $\mu = eu$, the magnitude of the dipole moment. This would be obtained from eqn. (4.8) if $e^{i\mathbf{k}\cdot\mathbf{u}}$ is set equal to unity and the dipole moment $(-e\mathbf{u})$ is treated as a scalar.

Using eqn. (4.19), (4.17) and (4.41) we can write the energy difference between the $2s$ and $2p$ states of hydrogen in the form,

$$
\Delta(E_{2s} - E_{2p}) \cong \hbar \int_{-\infty}^{\infty} d\xi \, \mathrm{Tr}\left\{\left(-\frac{\xi^2}{c^2}\mathcal{I} + (\nabla_r \, \nabla_r)\right)\int d^3 r'' G(\mathbf{r} - \mathbf{r}''; i\xi)\right.
$$
$$
\left. \times \left[\boldsymbol{\alpha}_{(2s)}(\mathbf{r}'' - \mathbf{r}'; i\xi) - \boldsymbol{\alpha}_{(2p)}(\mathbf{r}'' - \mathbf{r}'; i\xi)\right]\right\}_{\substack{r \to R \\ r' \to R}}
$$
$$
= -2\int_{-\infty}^{\infty} d\xi \left\{\left(\sum_m \frac{|\langle 2s| \, \mu \, |m\rangle|^2 \omega_{2s,m}}{\xi^2 + \omega_{2s,m}^2}\right.\right.
$$
$$
\left.\left. - \sum_{m'} \frac{|\langle 2p| \, \mu \, |m'\rangle|^2 \omega_{2p,m'}}{\xi^2 + \omega_{2p,m'}^2}\right)\right.
$$
$$
\left. \times \frac{1}{(2\pi)^2}\int \frac{d^3 k(3\xi^2/c^2 + k^2)}{\xi^2/c^2 + k^2}\right\}. \tag{4.42}
$$

Meaningful results can be extracted from the apparently divergent integral here in various ways. One way is to introduce a convergence factor $e^{ik\varepsilon}$ in the integral over k to get

$$
\lim_{\varepsilon \to 0+} \frac{1}{(2\pi)^3}\int_0^{\infty} \frac{4\pi k^2(3\xi^2/c^2 + k^2)}{\xi^2/c^2 + k^2} e^{ik\varepsilon} \, dk = -\frac{1}{2\pi}\left|\frac{\xi}{c}\right|^3. \tag{4.43}
$$

Thus, the integral over ξ can be written as,

$$
\Delta(E_{2s} - E_{2p}) = \frac{2}{\pi}\int_0^{\infty} d\xi \left(\frac{\xi}{c}\right)^3 \left(\sum_m \frac{|\langle 2s| \, \mu \, |m\rangle|^2 \omega_{2s,m}}{\xi^2 + \omega_{2s,m}^2}\right.
$$
$$
\left. - \sum_{m'} \frac{|\langle 2p| \, \mu \, |m'\rangle|^2 \omega_{2p,m'}}{\xi^2 + \omega_{2p,m'}^2}\right). \tag{4.44}
$$

This integral is divergent; but if we use a cut-off Λ at the upper limit, it can be written in the form,

$$
\Delta(E_{2s} - E_{2p}) = \frac{2}{\pi c^3}\left\{\frac{\Lambda^2}{2}\left(\sum_m |\langle 2s| \, \mu \, |m\rangle|^2 \omega_{2s,m}\right.\right.
$$
$$
\left.\left. - \sum_{m'} |\langle 2p| \, \mu \, |m'\rangle|^2 \omega_{2p,m'}\right)\right.
$$

$$-\frac{1}{2}\left(\sum_m |\langle 2s| \ \mu \ |m\rangle|^2 \omega^3_{2s,\,m} \ln \left|\frac{\Lambda^2 + \omega^2_{2s,\,m}}{\omega^2_{2s,\,m}}\right|\right.$$

$$\left.-\sum_{m'} |\langle 2p| \ \mu \ |m'\rangle|^2 \omega^3_{2p,\,m'} \ln \left|\frac{\Lambda^2 + \omega^2_{2p,\,m'}}{\omega^2_{2p,\,m'}}\right|\right)\biggr\}. \quad (4.45)$$

We can use the following sum rules (Power, 1966) to simplify eqn. (4.45):

$$\sum_m |\langle l| \ \mu \ |m\rangle|^2 \omega_{lm} = -\frac{e^2\hbar}{2m} \ ; \quad (4.46a)$$

$$\sum_m |\langle l| \ \mu \ |m\rangle|^2 \omega^3_{lm} = -\frac{e^2\hbar}{6m^2} \langle l| \ \nabla^2 V \ |l\rangle. \quad (4.46b)$$

Here $V = -(e^2/r)$, so that $\langle l| \ \nabla^2 V \ |l\rangle = (-4\pi e^2)|\phi_l(0)|^2$. Thus, the coefficient of the Λ^2 term in (4.45) cancels out, and since the logarithm is a slowly varying function of the argument, by setting

$$\sum_m |\langle 2s| \ \mu \ |m\rangle|^2 \omega^3_{2s,\,m} \cdot \frac{1}{2} \ln \left|\frac{\Lambda^2 + \omega^2_{2s,\,m}}{\omega^2_{2s,\,m}}\right|$$

$$= \sum_m |\langle 2s| \ \mu \ |m\rangle|^2 \omega^3_{2s,\,m} \ln (\omega_c/\bar{\omega}), \quad (4.47)$$

where ω_c is the cut-off frequency corresponding to Λ, and $\bar{\omega}$ is an average transition frequency, with $\omega_c \gg \bar{\omega}$, we get,

$$\Delta(E_{2s} - E_{2p}) \cong -\frac{4}{3}\left\{\left(\frac{e^2}{\hbar c}\right)^2 \left(\frac{\hbar}{mc}\right)^2 (\hbar c)|\phi_{2s}(0)|^2 \ln (\omega_c/\bar{\omega})\right\}. \quad (4.48)$$

This result for the Lamb shift was originally obtained by Bethe (1947) and by Welton (1948). In Bethe's calculation ω_c is the frequency corresponding to the Compton wavelength of the electron.

The detailed analysis given here is meant to emphasize the consistence between the formalism of dispersion self-energy of an atomic system developed here within a semi-classical framework, and the well-known quantum electrodynamic analysis of the same problem.

4.4 The effect of a material medium

If we consider the situation where there are N-molecules with polarizabilities $\alpha_1(\mathbf{r} - \mathbf{R}_1; \omega)$, $\alpha_2(\mathbf{r} - \mathbf{R}_2; \omega)$, ..., $\alpha_N(\mathbf{r} - \mathbf{R}_N; \omega)$, the expression for the interaction part of the dispersion energy of the system can be obtained exactly in the same manner as eqn. (2.40) for point molecules, in the form

$$\Delta E(N) = -\frac{\hbar}{4\pi i} \oint d\omega \ln \left(\frac{\{D_N(\omega)/D_0(\omega)\}}{\prod\limits_{j=1}^{N} \{D_j(\omega)/D_0(\omega)\}}\right), \quad (4.49)$$

where

$\{D_N(\omega)/D_0(\omega)\}$

$$
=
\begin{vmatrix}
\mathcal{I}+4\pi\mathcal{G}_1(\mathbf{R}_1,\mathbf{R}_1;\omega) & 4\pi\mathcal{G}_2(\mathbf{R}_1,\mathbf{R}_2;\omega) & \cdots & 4\pi\mathcal{G}_N(\mathbf{R}_1,\mathbf{R}_N;\omega) \\
4\pi\mathcal{G}_1(\mathbf{R}_2,\mathbf{R}_1;\omega) & \mathcal{I}+4\pi\mathcal{G}_2(\mathbf{R}_2,\mathbf{R}_2;\omega) & \cdots & \cdots \\
\cdot & \cdot & \cdots & \cdot \\
\cdot & \cdot & \cdots & \cdot \\
\cdot & \cdot & \cdots & \cdot \\
4\pi\mathcal{G}_1(\mathbf{R}_N,\mathbf{R}_1:\omega) & \cdots & \cdots & \mathcal{I}+4\pi\mathcal{G}_N(\mathbf{R}_N,\mathbf{R}_N;\omega)
\end{vmatrix}
$$

(4.50)

with

$$
\mathcal{G}_j(\mathbf{r},\mathbf{r}';\omega)=[\omega^2/c^2\mathcal{I}+(\nabla_r\,\nabla_r)]\int G(\mathbf{r}-\mathbf{r}'';\omega)\boldsymbol{\alpha}_j(\mathbf{r}''-\mathbf{r}';\omega)\,d^3\mathbf{r}''.
$$

(4.51)

We also have,

$$
\{D_j(\omega)/D_0(\omega)\}=|\mathcal{I}+4\pi\mathcal{G}_j(\mathbf{R}_j,\mathbf{R}_j;\omega)|.
$$

(4.52)

$\Delta E(N)$ in eqn. (4.49) is the mutual interaction energy of N molecules, since all the self-energies are subtracted off. It can be expanded in the same way as in eqn. (2.43), in terms of two-particle, three-particle, $\ldots$ dispersion interactions.

The total dispersion energy of the N-particle system including the self-energy terms is,

$$
\begin{aligned}
E(N) &= -\frac{\hbar}{4\pi i}\oint d\omega\,\ln\{D_N(\omega)/D_0(\omega)\} \\
&= -\frac{\hbar}{4\pi i}\oint d\omega\,\ln|\mathcal{I}+4\pi\mathcal{G}|,
\end{aligned}
$$

(4.53)

where $\mathcal{G}$ is a $3N\times 3N$ matrix, (ij)-th submatrix of which is $\mathcal{G}_j(\mathbf{R}_i,\mathbf{R}_j;\omega)$.

When the molecules constitute a material medium, in principle an analysis of the same sort as has been done in §1.7 can be made to work out the dispersion interaction between different parts of the medium, with $4\pi\mathcal{G}$ playing the role of $\boldsymbol{\alpha}\mathcal{T}$ in eqn. (1.80). The main difference will be in the fact that unlike the elements of $\mathcal{T}$ which diverge when two molecules get close, no such divergence will occur in $\mathcal{G}$. This absence of divergence has already been demonstrated in §4.2.

It is often useful to recast eqn. (4.53) in a form in which one needs to use the Green functions corresponding to the macroscopic fields in the dielectric media formed out of the molecules. This recasting, although not rigorously valid in all circumstances, can be done as follows. If we consider an external dipole source $\mathbf{P}_{\text{ext}}(\omega)$ at $\mathbf{R}_l$, the dipole moment induced by it on the molecule at $\mathbf{R}_j$ taking the screening effect of all the molecules as in eqn. (1.7), is given by*

$$\mathbf{P}_j(\omega) = -\{[\mathscr{I} + 4\pi\mathscr{G}]^{-1} 4\pi\mathscr{G}\}_{jl} \mathbf{P}_{\text{ext}}(\omega), \tag{4.54}$$

where

$$\mathbf{P}_j(\omega) = -\int d^3 r\, \mathbf{p}_j(\mathbf{r}, \omega) \tag{4.55a}$$

$$\mathbf{p}_j(\mathbf{r}, \omega) = \boldsymbol{\alpha}(\mathbf{r} - \mathbf{R}_j; \omega)\mathbf{E}(\mathbf{R}_j). \tag{4.55b}$$

But from a macroscopic point of view, the polarization of the j-th molecule can be obtained in the form

$$\mathbf{P}_j(\omega) = -4\pi\mathscr{G}_j^{(M)}(\mathbf{R}_l, \mathbf{R}_j; \omega)\, \mathbf{P}_{\text{ext}}(\omega), \tag{4.56}$$

where

$$\mathscr{G}_j^{(M)}(\mathbf{r}, \mathbf{r}'; \omega) = \int \mathscr{G}^{(M)}(\mathbf{r}, \mathbf{r}''; \omega)\boldsymbol{\alpha}_j(\mathbf{r}'' - \mathbf{r}'; \omega)\, d^3 r'' \tag{4.57}$$

and $4\pi\mathscr{G}^{(M)}(\mathbf{r}, \mathbf{r}'; \omega)$ is the dyadic Green function of the Maxwell equation in the material medium that connects the electric field at $\mathbf{r}$ with a dipole source at $\mathbf{r}'$. If we make the formal identification of the two equations (4.56) and (4.54), we obtain

$$\mathscr{G}^{(M)} = [\mathscr{I} + 4\pi\mathscr{G}]^{-1}\mathscr{G} \tag{4.58a}$$

or

$$[\mathscr{I} - 4\pi\mathscr{G}^{(M)}] = [\mathscr{I} + 4\pi\mathscr{G}]^{-1}. \tag{4.58b}$$

Here $\mathscr{G}^{(M)}$ is a $3N \times 3N$ matrix with the (ij)-th submatrix being $\mathscr{G}_i^{(m)}(\mathbf{R}_i, \mathbf{R}_j; \omega)$.

We thus get

$$\ln|\mathscr{I} + 4\pi\mathscr{G}| = -\ln|\mathscr{I} - 4\pi\mathscr{G}^{(M)}| \tag{4.59}$$

and eqn. (4.53) becomes

$$E(N) = \frac{\hbar}{4\pi i} \oint d\omega\, \ln|\mathscr{I} - 4\pi\mathscr{G}^{(M)}|. \tag{4.60}$$

We remark here that the identification of eqns. (4.54) and (4.56) implies that the electric field at $\mathbf{R}_j$ due to a dipole source $\mathbf{P}_l(\omega)$ at $\mathbf{R}_l$ is the same, both in the microscopic and macroscopic scales. This is true for

---

* We have tacitly assumed that all polarizability densities are the same, and $\alpha_j(\omega) = \alpha_l(\omega)$. The derivation of eqn. (4.54) is given in Appendix C. The validity of eqn. (4.60) is in fact independent of this assumption. The formulation could have been carried out entirely in terms of the macroscopic fields. See Appendix C.

dilute media for which the dielectric constant and molecular polarizabilities are related through the formula

$$\varepsilon(\omega) = 1 + 4\pi n\boldsymbol{\alpha}(\omega). \qquad (4.61)$$

In general, however, such an identification is not rigorous, and whereas eqn. (4.53) is exact, eqn. (4.60) is approximate, and the closeness of the approximation is essentially dependent on the extent to which the macroscopic field at a point in the medium (obtained by solving the Maxwell equation for material media) agrees with the field in the microscopic scale.

We shall use eqn. (4.60) to calculate the nature of the force field on a molecule near a dielectric interface in the next section.

4.5   The self-energy of a molecule near the interface of two media

If we expand the logarithm in (4.60), to leading order in the elements of $\mathscr{G}^{(M)}$ we get

$$E(N) = -\frac{\hbar}{i} \oint d\omega \sum_{j=1}^{N} \text{Tr } \mathscr{G}_j^{(M)}(\mathbf{R}_j, \mathbf{R}_j; \omega). \qquad (4.62)$$

To this order of approximation $E(N)$ is clearly the sum of the self-energies of the molecules, the self-energy of the l-th molecule being

$$E_s(l) = -\frac{\hbar}{i} \oint d\omega \text{ Tr } \mathscr{G}^{(M)}(\mathbf{R}_l, \mathbf{R}_l; \omega). \qquad (4.63)$$

It must be noted that if we include the traces of the higher powers, eqn. (4.62) is not necessarily a simple power series in powers of the polarizabilities of the molecules, as one would get by making the same sort of expansion for eqn. (4.53). Each term in the series for (4.62) would be a complicated function of the polarizabilities, since each term would involve the dielectric constant which in a dense medium depends on the molecular polarizabilities in a complicated way. We shall return to this point later.

Equation (4.63) can be used to estimate the energy of a single molecule near the interface of two dielectric media. We consider only the non-retarded limit. The geometry of the situation is given in Fig. 4.2. $\mathscr{G}^{(M)}(\mathbf{r}, \mathbf{r}'; \omega)$ is evaluated from the equations

$$\mathscr{G}^{(M)}(\mathbf{r}, \mathbf{r}') = -\nabla_r \nabla_{r'} \int G^{(M)}(\mathbf{r}, \mathbf{r}'')\boldsymbol{\alpha}(\mathbf{r}'' - \mathbf{r}'; \omega) \, d^3r \qquad (4.64)$$

with

$$\nabla^2 G^{(M)}(\mathbf{r}, \mathbf{r}') = \delta(\mathbf{r} - \mathbf{r}') \qquad (4.65)$$

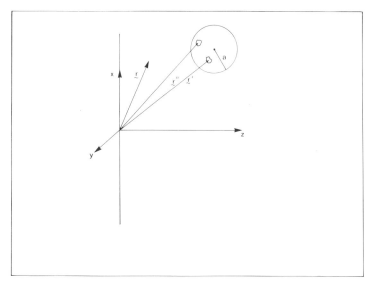

FIG. 4.2. Molecule near an interface, indicating the geometry involved in eqn. (4.64). The dielectric constants are ε_1 to the left and ε_2 to the right of the interface.

and with the boundary conditions that $G^{(M)}$ and $\varepsilon(\partial G^{(M)}/\partial z)$ be continuous across the boundary. Explicitly, we have

$$G^{(M)}(\mathbf{r}, \mathbf{r}') = -\frac{1}{4\pi}\left(\frac{\theta(z')}{\varepsilon_2} + \frac{\theta(-z')}{\varepsilon_1}\right)\left(\frac{1}{|\mathbf{r}-\mathbf{r}'|}\right.$$

$$\left. - \frac{\mathrm{sgn}\,(z')\Delta_{12}}{\{(x-x')^2+(y-y')^2+(|z|+|z'|)^2\}^{1/2}}\right), \quad (4.66)$$

where $\mathrm{sgn}\,(z')$ and $\theta(z')$ are the signum and step functions of z' and

$$\Delta_{ij} = (\varepsilon_i - \varepsilon_j)/(\varepsilon_i + \varepsilon_j). \quad (4.67)$$

If we assume a Gaussian form for $\boldsymbol{\alpha}(\mathbf{r}-\mathbf{R}; \omega)$ of the sort given in (4.20) and (4.21), using eqns. (4.64) and (4.66) we get

$$\mathrm{Tr}\,\mathscr{G}^{(M)}(\mathbf{R}, \mathbf{R}; \omega) = \frac{\alpha(\omega)}{2\pi^{3/2}a^3}\left\{2\left(\frac{\theta(z)}{\varepsilon_2} + \frac{\theta(-z)}{\varepsilon_1}\right)\right.$$

$$-\Delta_{12}\left(\frac{\theta(z)}{\varepsilon_2} - \frac{\theta(-z)}{\varepsilon_1}\right)\left(\exp\,(-z^2/a^2) - \frac{\sqrt{\pi}|z|}{a}\,\mathrm{erfc}\left(\frac{|z|}{a}\right)\right.$$

$$\left.\left. + \frac{a^3}{2|z|^3}\int_0^{|z|/a}(\mathrm{e}^{-t^2} - \mathrm{e}^{-z^2/a^2})\,\mathrm{d}t\right)\right\}. \quad (4.68)$$

In this expression z is the z-component of $\mathbf{R}$, the position of the molecule. Substituting this in (4.63) and with $\omega \equiv i\xi$, we get the dispersion self-energy of a molecule as a function of its distance z from the interface in the form

$$
E_s(z) = \frac{\hbar}{\pi^{3/2} a^3} \int_0^\infty d\xi\, \alpha(i\xi) \left\{ 2\left(\frac{\theta(z)}{\varepsilon_2} + \frac{\theta(-z)}{\varepsilon_1}\right)\right.
$$

$$
- \Delta_{12}(i\xi)\left(\frac{\theta(z)}{\varepsilon_2} - \frac{\theta(-z)}{\varepsilon_1}\right)\left(\exp\left(-z^2/a^2\right)\right.
$$

$$
\left.\left. - \frac{\sqrt{\pi}|z|}{a}\operatorname{erfc}\left(\frac{|z|}{a}\right) + \frac{a^3}{2|z|^3}\int_0^{|z|/a}(e^{-t^2} - e^{-z^2/a^2})\,dt\right)\right\}. \quad (4.69)
$$

Here the dielectric constants ε_1 and ε_2 depend on the frequency.

This expression for the self-energy of the molecule at position $\mathbf{R}$ is particularly significant, as it gives the force field operating on it arising out of the dispersion interaction between this molecule and the others in the two media. For the various limiting values of z we have,

$$
E_s(z) = \frac{2\hbar}{\pi^{3/2} a^3 \varepsilon_1} \int_0^\infty d\xi\, \alpha(i\xi); \qquad\qquad z \to -\infty
$$

$$
= \frac{2\hbar}{\pi^{3/2} a^3 \varepsilon_2} \int_0^\infty d\xi\, \alpha(i\xi); \qquad\qquad z \to +\infty
$$

$$
= \frac{2\hbar}{\pi^{3/2} a^3} \int_0^\infty d\xi\, \alpha(i\xi)\left\{\frac{1}{2}\left(\frac{1}{\varepsilon_1} + \frac{1}{\varepsilon_2}\right) - \frac{\Delta_{12}}{3}\left(\frac{1}{\varepsilon_2} - \frac{1}{\varepsilon_1}\right)\right\}; \qquad z = 0.
$$

$$
\tag{4.70}
$$

Furthermore, the force on the molecule is given by (for $|z| \gg a$),

$$
\mathscr{F}(z) = -\frac{\partial E_s(z)}{\partial z}
$$

$$
\cong -\frac{3\hbar}{4\pi|z|^4}\operatorname{sgn}(z)\int_0^\infty d\xi\, \alpha(i\xi)\, \Delta_{12}\left(\frac{\theta(z)}{\varepsilon_2} - \frac{\theta(-z)}{\varepsilon_1}\right). \quad (4.71)
$$

This $1/|z|^4$ dependence of the dispersion force between a molecule and a dielectric interface is well known and has recently been rederived by Israelachvili (1972) from a different point of view. For $z \approx a$ there is a deviation of the force law from the $1/|z|^4$ form, and as $|z| \to 0$ from either side it tends to a finite limit, unlike the divergent result one would get in extrapolating from the $1/|z|^4$ force law.

The z^{-4} form seems to be a good representation of the force law for $|z| \gtrsim 2a$, so that if we compute the Lifshitz non-retarded force per unit area between two dielectric slabs a distance l apart from this $1/z^4$ force law, we can expect the resulting $1/l^3$ law to be valid up to a lower limit for l of the order of the sizes of the molecules.

The main approximation in the analysis given above arises out of the assumption of an infinitely sharp interface between the two media, even when the molecules have finite size. This somewhat inconsistent assumption does not alter the main conclusions, although the results for $z \to 0$ have to be regarded as somewhat qualitative. The important point that emerges is that in an inhomogeneous situation $E_s(\mathbf{R})$ of a molecule depends on $\mathbf{R}$. In fact, $E_s(\mathbf{R})$ defines a sort of "dispersion potential" that the molecule experiences in the presence of the other molecules. It is analogous to the density-dependent potential used in the theory of the inhomogeneous electron gas (Kohn and Sham, 1965; Beattie, Stoddart and March, 1971).

4.6 The surface energy via self-energy

To define a surface energy based on the above concept of self-energy, let us take the media 1 and 2 as consisting of a uniform distribution of isotropic molecules of radii a_1 and a_2, polarizabilities α_1 and α_2 and the densities n_1 and n_2. The presence of the interface alters the self-energy of each molecule from what it would be in a homogeneous system. From the analysis of the preceding section we note that the atoms in a layer of thickness dz in medium 1 at a distance $|z|$ from the interface would have a change in their energy per unit area given by

$$n_1 \Delta E_1(z) = \frac{\hbar n_1}{\pi^{3/2} a_1^3} \int_0^\infty d\xi \alpha_1(i\xi) \frac{\Delta_{12}}{\varepsilon_1} \left\{ \exp\left(-z^2/a_1^2\right) \right.$$

$$\left. - \frac{\sqrt{\pi}|z|}{a_1} \text{erfc}\left(\frac{|z|}{a_1}\right) + \frac{a_1^3}{2|z|^3} \int_0^{|z|/a_1} (e^{-t^2} - e^{-z^2/a_1^2}) \, dt \right\}, \quad (4.72)$$

where

$$\Delta E_1(z) = E_s(z) - E_s(-\infty). \quad (4.73)$$

Integrating over z from $(-\infty)$ to 0 we get the total change in the energy of medium 1 per unit area in the form,

$$E_{s(1)} = +(\hbar n_1/2\pi a_1^2) \int_0^\infty d\xi [\alpha_1(i\xi)/\varepsilon_1] \Delta_{12}. \quad (4.74)$$

Similarly, from the other side we get

$$E_{s(2)} = -(\hbar n_2/2\pi a_2^2) \int_0^\infty d\xi [\alpha_2(i\xi)/\varepsilon_2] \Delta_{12}. \tag{4.75}$$

Thus the final expression for the surface energy becomes

$$E_{\text{Surface}} = +\frac{\hbar}{2\pi} \int_0^\infty d\xi \, \Delta_{12} \left(\frac{n_1\alpha_1}{\varepsilon_1 a_1^2} - \frac{n_2\alpha_2}{\varepsilon_2 a_2^2} \right). \tag{4.76}$$

The result is only an approximation, since we have retained only leading terms $0(\mathscr{G}^{(M)})$ in the expansion of (4.60). Nonetheless the expression for the surface energy contains the essential physical aspects of the dispersion contribution to the surface energy, even though the interface is regarded as infinitely sharp. A more realistic approach would be to assume that there is a gradual variation in the density of the molecules across the interface; but this refinement will not alter the main conclusions, particularly the way in which the size of the molecules enters into the expression for the surface energy.

This theory of surface energy in non-polar media has been developed by Mahanty and Ninham (1974). We shall briefly discuss here the implication of this theory in relation to some other theories of surface energy.

For most non-polar dielectrics $(\varepsilon_j - 1)/4\pi n\alpha_j < 1$. Then

$$(\varepsilon_j)^{\pm 1} \approx 1 \pm 4\pi n_j \alpha_j. \tag{4.77}$$

Also, if the molecules in the two media have one principal absorption frequency each, then $\alpha_j(\omega)$ can be written in the form,

$$\alpha_j(\omega) = \alpha_j(0)/(1 - \omega^2/\omega_j^2). \tag{4.78}$$

Then to leading order in $\alpha_j(0)$, after carrying out the frequency integration in eqn. (4.76) we get

$$E_{\text{Surface}} = +\frac{\hbar\pi}{2} \left\{ \frac{n_1^2\alpha_1^2(0)\omega_1}{a_1^2} + \frac{n_2^2\alpha_2^2(0)\omega_2}{a_2^2} \right.$$
$$\left. - \frac{2n_1 n_2 \alpha_1(0)\alpha_2(0)\omega_1\omega_2}{(\omega_1 + \omega_2)} \left(\frac{1}{a_1^2} + \frac{1}{a_2^2} \right) \right\}. \tag{4.79}$$

In the case of close packing we may assume that a_j is the distance between two neighbouring molecules in medium j, i.e., the distance of separation is of the order of the size of the molecule.* In that case eqn. (4.79) reduces further to a form essentially equivalent to that given by

---

* With a Gaussian form factor, a_j could be related to the usual "hard core" radius through the use of the Rayleigh or a similar criterion for resolution of two intensity functions.

Fowkes (1964, 1968) for the interfacial energy of two dielectric media, obtained by pairwise summation of the London interaction between the molecules. If the interface separates the medium from vacuum (i.e., $n_2 = 0$), (4.79) reduces to

$$E_{\text{Surface}} = \pi/2\hbar n_1^2[\alpha_1^2(0)\omega_1/a_1^2].\qquad(4.80)$$

This last result is in good numerical agreement with corresponding expressions given by Fowkes if a_1 is interpreted as the distance between neighbouring molecules in the medium. For detailed calculations for real systems see Fowkes (1964, 1968) and Israelachvili (1973).

If $a_1 \approx a_2$, i.e., if the atoms of the media have nearly the same size, eqn. (4.76) can be rewritten as

$$E_{\text{Surface}} \approx \frac{\hbar}{8\pi^2 a^2} \int_0^\infty \Delta_{12}\left(\frac{1}{\varepsilon_2} - \frac{1}{\varepsilon_1}\right) d\xi.\qquad(4.81)$$

This form is similar to an expression for the surface energy obtained by Craig (1972). (See also Mitchell and Richmond (1973, 1974).) In his formulation Craig obtained the surface energy in terms of a two-dimensional integral over the wave-vector $k_\parallel$ of a combination of wave–vector dependent dielectric constants in the same form as in the integrand of (4.81). His formalism involving k-dependent dielectric constants is incorrect, but is all right when applied to k-independent dielectrics. In that case a cut-off k_c has to be used in the $k_\parallel$ integration and that would lead to a factor like $1/a^2$ where $a \approx 1/k_c$. There is, in fact, no reason why the cut-offs should be the same for the two dielectric media, and in any case, the cut-offs can be meaningful only if related to the sizes of the molecules in the two media. Hence eqn. (4.76) is of more general validity than that obtainable from Craig's results for k-independent dielectric media.

While the effects of temperature could be included by assigning an oscillator free energy

$$F_\omega = k_B T \ln\left[2\sinh(\hbar\omega/2k_B T)\right]$$

to each mode instead of a zero-point energy $\frac{1}{2}\hbar\omega$ these effects are normally of little importance as far as the dispersion part of the interaction is concerned.* For polar media, an extension of the method developed here to take account of orientation effects of permanent dipoles near a surface must be followed. One way of effecting this extension will be indicated in Chapter 7.

---

* See Israelachvili (1973) who shows that the actual temperature dependence of surface free energies for liquid hydrocarbons is given correctly by this generalization.

4.7 Interaction between macroscopic bodies

The interaction between macroscopic bodies (necessarily) constituted of molecules of finite size can be studied in the same sort of formalism developed in §1.7 (Mahanty and Ninham, 1974). Starting with eqn. (4.53), if we consider the situation where we have two bodies A and B separated by a vacuum we can write the determinant $|\mathscr{I} + 4\pi\mathscr{G}|$ as

$$|\mathscr{I} + 4\pi\mathscr{G}| = \begin{vmatrix} \mathscr{I} + 4\pi\mathscr{G}_A & 4\pi\mathscr{G}_{AB} \\ 4\pi\mathscr{G}_{BA} & \mathscr{I} + 4\pi\mathscr{G}_B \end{vmatrix}. \qquad (4.82)$$

Here $\mathscr{G}_A$ and $\mathscr{G}_B$ refer to the molecules within A and within B, whereas $\mathscr{G}_{AB}$ and $\mathscr{G}_{BA}$ each refer to the interaction between a molecule represented by the first subscript with one in the body represented by the second. If N_A and N_B represent the number of molecules in A and B, the dimensions of the submatrices in eqn. (4.82) are: $\mathscr{G}_A$ is $3N_A \times 3N_A$, $\mathscr{G}_{AB}$ is $3N_A \times 3N_B$, $\mathscr{G}_{BA}$ is $3N_B \times 3N_A$ and $\mathscr{G}_{BB}$ is $3N_B \times 3N_B$. The (jl)-th element of these matrices, which are 3×3 submatrices are given by

$$\mathscr{G}_{AB}(j, l)\Big|_{\substack{j \in A \\ l \in B}} = -[\omega^2/c^2\mathscr{I} - \nabla_r\nabla_{r'}] \int G(\mathbf{r} - \mathbf{r}''; \omega)$$

$$\times \alpha_l^{(B)}(\mathbf{r}'' - \mathbf{r}'; \omega)\, \mathrm{d}^3 r''\Big|_{\substack{r \to R_j \\ r' \to R_l}}, \qquad (4.83)$$

where $\mathbf{R}_j$ and $\mathbf{R}_l$ are the positions of the j-th and l-th molecules and similarly for

$$\mathscr{G}_A(j, l)\Big|_{\substack{j \in A \\ l \in A}}, \quad \mathscr{G}_{BA}(j, l)\Big|_{\substack{j \in B \\ l \in A}} \quad \text{and} \quad \mathscr{G}_B(j, l)\Big|_{\substack{j \in B \\ l \in B}}.$$

With a little algebra, we can write

$$\ln|I + 4\pi\mathscr{G}| = \ln|\mathscr{I} + 4\pi\mathscr{G}_A| + \ln|\mathscr{I} + 4\pi\mathscr{G}_B|$$
$$+ \ln|\mathscr{I} - \{(\mathscr{I} + 4\pi\mathscr{G}_A)^{-1}4\pi\mathscr{G}_{AB}\}\{(\mathscr{I} + 4\pi\mathscr{G}_B)^{-1}4\pi\mathscr{G}_{BA}\}|. \quad (4.84)$$

As in §1.7, the significance of the matrix $\{(\mathscr{I} + 4\pi\mathscr{G}_A)^{-1}4\pi\mathscr{G}_{AB}\}$ can be established as follows. Following the reasoning associated with eqn. (4.54) we can interpret the (jl)-th element of this matrix as the one which connects the polarization of the l-th molecule, $l \in B$, with the field at the j-th molecule, $j \in A$, taking the induced polarization of all the molecules in A into account. Similarly, the (jl)-th element of $\{(\mathscr{I} + 4\pi\mathscr{G}_B)^{-1}4\pi\mathscr{G}_{BA}\}$ connects the polarization of the l-th molecule, $l \in A$, with the field at the j-th molecule, $j \in B$, taking into account the induced polarization of all the molecules in B. For various geometries of the bodies A and B, in principle at least, the various elements of these matrices can be evaluated by the summation procedure stated in §1.7. We can also define macroscopic Green functions

$$\mathscr{G}_{AB}^{(M)} \equiv (\mathscr{I} + 4\pi\mathscr{G}_A)^{-1}\mathscr{G}_{AB} \tag{4.85a}$$

$$\mathscr{G}_{BA}^{(M)} \equiv (\mathscr{I} + 4\pi\mathscr{G}_B)^{-1}\mathscr{G}_{BA} \tag{4.85b}$$

and assume that their elements are obtainable from solutions of Maxwell's equations with appropriate boundary conditions, treating A and B as dielectric continua, and then using eqn. (4.57) to take account of molecular size. In other words,

$$\mathscr{G}_{AB}^{(M)}(\mathbf{r},\mathbf{r}';\omega) \equiv -\int \nabla_r \nabla_{r'} G^{(M)}(\mathbf{r},\mathbf{r}'';\omega)\boldsymbol{\alpha}^{(B)}(\mathbf{r}''-\mathbf{r}';\omega)\,\mathrm{d}^3r'', \tag{4.86}$$

where $\mathscr{G}_{AB}^{(M)}(\mathbf{r},\mathbf{r}';\omega)$ is the dyadic Green function of the Maxwell equation that connects the field at $r \in A$ due to a source at $r' \in B$, ignoring the effect of the other molecules in B, but taking account of the dielectric properties of A. A similar equation can be written for the elements of $\mathscr{G}_{BA}^{(M)}(\mathbf{r},\mathbf{r}';\omega)$.

In eqn. (4.84), the first two terms on the right hand side, when used in eqn. (4.53) give (formally) the self-energies of the bodies A and B. They include both the bulk and surface energies of the isolated bodies. The last term gives the interaction energy. Using eqns. (4.85a) and (4.85b) we obtain

$$E_{AB} = -\frac{\hbar}{4\pi i}\oint \mathrm{d}\omega \ln |\mathscr{I} - (4\pi)^2\mathscr{G}_{AB}^{(M)}\mathscr{G}_{BA}^{(M)}|. \tag{4.87}$$

This equation provides a satisfactory basis for a theory of dispersion interaction of macroscopic bodies. An important implication of the form of the interaction energy given in eqn. (4.87) is that for point molecules E_{AB} can be derived by considering the field equations due to mutually induced surface currents on A and B. This follows from the fact that the effect of all the sources in B at a field point in A (taking the polarization of all the molecules in A into account) can be obtained equivalently from a surface distribution of currents on B, and similarly for the body A. The secular determinant which follows from the equations governing the field at the surfaces of A and B due to mutually induced surface currents is identical to the argument of the logarithm in eqn. (4.87). This aspect has been indicated in §2.4.

The result (4.87) is identical in form to the one first given by Renne (1971), but the molecular model here is not restricted to the point harmonic oscillator form as in Renne's analysis, and molecular size is taken into account explicitly. We now illustrate its use in the derivation of the Lifshitz force between two semi-infinite slabs (Fig. 4.3) and consider only the non-retarded limit.

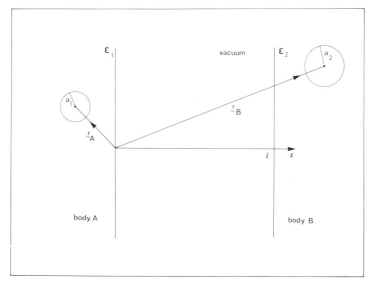

FIG. 4.3. Two molecules embedded in two semi-infinite parallel slabs separated by vacuum.

The form of the dyadic Green function needed for this problem given by (4.86) follows from eqn. (4.66). For a source molecule of B in vacuum $(z'>0)$ and field point at a molecule of $A(z<0)$, (4.66) reduces to

$$G^{(M)}(\mathbf{r},\mathbf{r}') = -\frac{1}{2\pi(\varepsilon_1+1)|\mathbf{r}-\mathbf{r}''|} = -\frac{2}{(2\pi)^3(\varepsilon_1+1)}\int\frac{\mathrm{d}^3k}{k^2}e^{i\mathbf{k}.(\mathbf{r}-\mathbf{r}'')}. \quad (4.88)$$

Hence, provided that $l>a_1+a_2$ [i.e., essentially, no molecules of B spill over to overlap with molecules A], and assuming a Gaussian form factor for the molecules as indicated in eqns. (4.20) and (4.21); viz.:

$$\begin{aligned}\boldsymbol{\alpha}(\mathbf{r}''-\mathbf{r}';\omega) &= \mathscr{I}\frac{\alpha(\omega)}{\pi^{3/2}a^3}e^{-(\mathbf{r}''-\mathbf{r}')^2/a^2}\\ &= \mathscr{I}\frac{\alpha(\omega)}{(2\pi)^3}\int\mathrm{d}^3k\,e^{i\mathbf{k}.(\mathbf{r}''-\mathbf{r}')}e^{-k^2a^2/4},\end{aligned}\quad (4.89)$$

we have from eqn. (4.86)

$$\begin{aligned}\mathscr{G}_{AB}^{(M)}(\mathbf{r}_A,\mathbf{r}_B) &= -\nabla_{r_A}\nabla_{r_B}\int G^{(M)}(\mathbf{r}_A,\mathbf{r}'';\omega)\boldsymbol{\alpha}^{(B)}(\mathbf{r}''-\mathbf{r}_B;\omega)\,\mathrm{d}^3r''\\ &= -\nabla_{r_A}\nabla_{r_B}\left(\frac{2\alpha_2(\omega)}{(2\pi)^6(\varepsilon_1+1)}\right)\mathscr{I}\int\frac{\mathrm{d}^3k}{k^2}\int\mathrm{d}^3k'\exp\left(-k'^2a_2^2/4\right)\end{aligned}$$

$$\times \int d^3 r'' e^{i\mathbf{k}\cdot(\mathbf{r}_A - \mathbf{r}'')} e^{i\mathbf{k}'\cdot(\mathbf{r}'' - \mathbf{r}_B)}\Big)$$

$$= -\frac{2\alpha_2(\omega)}{(2\pi)^3(\varepsilon_1 + 1)} \int d^3 k \, \frac{(\mathbf{k}\,\mathbf{k})}{k^2} e^{i\mathbf{k}\cdot(\mathbf{r}_A - \mathbf{r}_B)} e^{-k^2 a_2^2/4}.$$

(4.90a)

Similarly

$$[\mathscr{G}_{BA}^{(M)}(\mathbf{r}_B, \mathbf{r}_A)]_{jl} = -\frac{2\alpha_1(\omega)}{(2\pi)^3(\varepsilon_2 + 1)} \int \frac{d^3 k}{k^2} (\mathbf{k}\,\mathbf{k}) e^{i\mathbf{k}\cdot(\mathbf{r}_B - \mathbf{r}_A)} e^{-k^2 a_1^2/4}. \quad (4.90b)$$

A typical (3×3) submatrix of the product $\mathscr{G}_{AB}^{(M)}\mathscr{G}_{BA}^{(M)}$ is

$$(\mathscr{G}_{AB}^{(M)}\mathscr{G}_{BA}^{(M)})_{jl} = \frac{1}{(2\pi)^8} \frac{\Delta_{10}\Delta_{20}}{n_1 n_2} \int \frac{d^3 k \, d^3 k' \, d^3 r_B}{k^2 k'^2} \, n_2 e^{i\mathbf{k}\cdot(\mathbf{r}_{A^j} - \mathbf{r}_B)}$$

$$\times e^{i\mathbf{k}'\cdot(\mathbf{r}_B - \mathbf{r}_{A^l})} e^{-(k^2 a_1^2 + k'^2 a_2^2)/4} (\mathbf{k}\,\mathbf{k})(\mathbf{k}'\,\mathbf{k}'). \quad (4.91)$$

We have used the relation*

$$\alpha_j(\omega) = (\varepsilon_j - 1)/4\pi n_j, \qquad j = 1, 2 \quad (4.92)$$

where n_j is the density of molecules in the j-th medium, and

$$\Delta_{ij} = (\varepsilon_i - \varepsilon_j)/(\varepsilon_i + \varepsilon_j) \quad \text{with} \quad \varepsilon_0 \equiv \varepsilon_{\text{vacuum}} = 1. \quad (4.93)$$

For point molecules (i.e., $a_j = 0$, $j = 1, 2$), eqn. (4.91) reduces to a simple form. Carrying out the integration over $\mathbf{r}_B$ we get

$$(\mathscr{G}_{AB}^{(M)}\mathscr{G}_{BA}^{(M)})_{jl} = \frac{\Delta_{10}\Delta_{20}\pi^2}{2n_1(2\pi)^6} \int \frac{d^2\kappa}{\kappa^3} \{2\kappa^2 \mathscr{M}(\kappa)\} e^{i\mathbf{\kappa}\cdot(\mathbf{\rho}_{A^j} - \mathbf{\rho}_{A^l})} e^{-\kappa(2l - z_{A^j} - z_{A^l})}, \quad (4.94)$$

where the tensor $\mathscr{M}(\kappa)$ is

$$\mathscr{M}(\kappa) = (\mathbf{\kappa} - i\kappa\hat{e}_3)(\mathbf{\kappa} + i\kappa\hat{e}_3), \quad (4.95)$$

$\mathbf{\rho}, \mathbf{\kappa}$ are the projections of $\mathbf{r}, \mathbf{k}$ in the $(x - y)$ plane, and $\hat{e}_3$ is the unit vector in the z direction. The tensor $\mathscr{M}(\mathbf{\kappa})$ has the properties

$$\mathscr{M}(\mathbf{\kappa})\mathscr{M}(\mathbf{\kappa}) = 2\kappa^2 \mathscr{M}(\kappa); \qquad \text{Tr } \mathscr{M}(\kappa) = 2\kappa^2. \quad (4.96)$$

Thus, per unit area, we have

$$\text{Tr } (\mathscr{G}_{AB}^{(M)}\mathscr{G}_{BA}^{(M)}) = \left(\frac{\Delta_{10}\Delta_{20}}{(4\pi)^2}\right) \frac{1}{4\pi^2} \int d^2\kappa e^{-2\kappa l} \quad (4.97a)$$

$$\text{Tr } (\mathscr{G}_{AB}^{(M)}\mathscr{G}_{BA}^{(M)}\mathscr{G}_{AB}^{(M)}\mathscr{G}_{BA}^{(M)}) = \left(\frac{\Delta_{10}\Delta_{20}}{(4\pi)^2}\right)^2 \frac{1}{4\pi^2} \int d^2\kappa e^{-4\kappa l} \quad (4.97b)$$

---

* This assumption is not essential for the validity of eqn. (4.99) which is generally valid. (See Appendix C for proof.)

and so on. Hence

$$\ln |\mathcal{I} - (4\pi)^2 \mathcal{G}_{AB}^{(M)} \mathcal{G}_{BA}^{(M)}| = -\sum_{n=1}^{\infty} \frac{(4\pi)^{2n}}{n} \mathrm{Tr}\, (\mathcal{G}_{AB}^{(M)} \mathcal{G}_{BA}^{(M)})^n$$

$$= \frac{1}{4\pi^2} \int d^2\kappa \ln (1 - \Delta_{10} \Delta_{20} e^{-2\kappa l}). \quad (4.98)$$

Substitution of this expression into (4.87) gives the Lifshitz interaction energy

$$E_{AB} = \frac{\hbar}{4\pi^2} \int_0^{\infty} \kappa\, d\kappa \int_0^{\infty} d\xi \ln (1 - \Delta_{10}(i\xi) \Delta_{20}(i\xi) e^{-2\kappa l}). \quad (4.99)$$

When the size of the molecules is taken into account the computation becomes rather involved. The nature of the main result can be established by analysing the leading term only of E_{AB},

$$E_{AB} \approx \frac{\hbar}{4\pi i} \oint d\omega\, \mathrm{Tr}\, (16\pi^2 \mathcal{G}_{AB}^{(M)} \mathcal{G}_{BA}^{(M)})$$

$$= \frac{\hbar}{4\pi i} \oint d\omega \frac{\Delta_{10} \Delta_{20}}{(2\pi)^8} 16\pi^2 \int \frac{d^3k\, d^3k'}{k^2 k'^2} d^3r_B\, d^3r_A$$

$$\times e^{i(\mathbf{k}-\mathbf{k}')\cdot(r_A - r_B)} e^{-(k^2 a_1^2 + k'^2 a_2^2)/4} (\mathbf{k} \cdot \mathbf{k}')^2. \quad (4.100)$$

This reduces to the form (4.34) already obtained by pairwise summation of London interaction energy of the constituent molecules, taking their size into account, if we assume $\Delta_{j0}/4\pi = \alpha_j n_j$. We thus note that the divergence of Lifshitz theory at zero separation of macroscopic bodies is removed.

Strictly speaking eqns. (4.34) and (4.100) cannot be extrapolated down to zero separation. This is apart from assumptions already made concerning the relation between macroscopic and microscopic Green functions, the assumed planar profile, and the approximation of replacing sums by integrals in evaluating the required traces. The special form of the macroscopic Green function (4.66) which we have used [eqn. (4.88)] in evaluating $\mathcal{G}_{AB}^{(M)}$ is that simple form of (4.66) which holds for a situation where source and field points belong to different media [cf. Fig. 4.3]. In fact at $l \to 0$, the complete form of (4.66) must be used if the molecules can overlap, and the calculation, even to lowest order in polarizability becomes much more complicated. Whether a model in which $l < a_1 + a_2$ is meaningful or not is an open question—for at least some situations it may well provide a simple and useful model for chemical binding in condensed media interactions. But in any event the interaction energy is in fact always finite at zero separation.

Effect of an intervening substance

If the two bodies A and B are completely embedded in a medium C, the interaction energy between them will be affected by the dielectric constant of the medium C. The simplest way of taking this into consideration is to work out the energy for the medium C without A and B, and to evaluate the change in its energy due to the insertion of A and B. The insertion of A and B implies replacement of the molecules of C by molecules of A and B in the region occupied by the latter. This entails a change of polarizability of the region, by an amount

$$(n_A\alpha_A - n_C\alpha_C) = \frac{\varepsilon_A - \varepsilon_C}{4\pi} \qquad (4.101)$$

if we use eqn. (4.92). We can thus take account of the intervening medium by replacing the molecular polarizability α_A of A by an effective polarizability

$$\alpha_A^* = \alpha_A - \frac{n_C}{n_A}\alpha_C = \frac{\varepsilon_A - \varepsilon_C}{4\pi n_A}. \qquad (4.102)$$

Similarly, α_B can be replaced by

$$\alpha_B^* = \alpha_B - \frac{n_C}{n_B}\alpha_C = \frac{\varepsilon_B - \varepsilon_C}{4\pi n_B}. \qquad (4.103)$$

When these effective polarizabilities are used in the Green functions occurring in eqns. (4.90), (4.91) and in the subsequent steps we get the interaction energy in the presence of the medium C. The modification of the formula of eqn. (4.99) for the interaction energy due to the presence of a dielectric medium will be the replacement of $\Delta_{10}\,\Delta_{20}$ by $\Delta_{13}\,\Delta_{23}$, where $\Delta_{j3} = (\varepsilon_j - \varepsilon_3)/(\varepsilon_j + \varepsilon_3)$.

4.8 Surface and interaction energy

An interesting question is the connection between the Lifshitz interaction energy between two bodies and the surface energies of each taken in isolation. To examine this question, we go back to eqn. (4.60), and consider first a situation where we have only the body A. Then $\ln|\mathcal{G} + 4\pi\mathcal{G}_A|$ in (4.60) would give us the total dispersion energy of body A which is the sum of the bulk and surface energies. The definition of surface energy is the difference between the energy of A alone, and that of A if it were completely embedded in a medium of the same sort, i.e., if the surface were absent. The latter can be obtained by considering two bodies A and A' in contact, the latter being a medium consisting of molecules of the same type as A filling all space excluding A. By an analysis similar to what is done in eqn. (4.84) the total dispersion energy of the bulk medium consisting of A and A' will be obtained from

$$\ln |\mathscr{I} + 4\pi\mathscr{G}_A| + \ln |\mathscr{I} + 4\pi\mathscr{G}_{A'}| + \ln |\mathscr{I} - (4\pi)^2\mathscr{G}_{AA}^{(M)}\mathscr{G}_{A'A}^{(M)}|.$$

Thus the energy of A embedded in the homogeneous medium is obtained from

$$\ln |\mathscr{I} + 4\pi\mathscr{G}_A| + \tfrac{1}{2}\ln |\mathscr{I} - (4\pi)^2\mathscr{G}_{AA}^{(M)}\mathscr{G}_{A'A}^{(M)}|,$$

and the surface energy of A alone is obtained from

$$\{-\tfrac{1}{2}\ln |\mathscr{I} - (4\pi)^2\mathscr{G}_{AA}^{(M)}\mathscr{G}_{A'A}^{(M)}|\}.$$

Similarly the surface energy of B arises out of a term

$$\{-\tfrac{1}{2}\ln |\mathscr{I} - (4\pi)^2\mathscr{G}_{BB}^{(M)}\mathscr{G}_{B'B}^{(M)}|\},$$

where B' is of the same material as B filling up all space excluding B.

The surface energy of A and B when they are separated by a distance l is the difference between the total energy of the system and the sum of the energies of A and B when they are embedded respectively in media of the same sort. Thus the surface energy would be obtained from the expression

$$\ln\{|\mathscr{I} + 4\pi\mathscr{G}_A||\mathscr{I} + 4\pi\mathscr{G}_B||\mathscr{I} - (4\pi)^2\mathscr{G}_{AB}^{(M)}\mathscr{G}_{BA}^{(M)}|\} - \ln |\mathscr{I} + 4\pi\mathscr{G}_A|$$

$$- \ln |\mathscr{I} + 4\pi\mathscr{G}_B| - \tfrac{1}{2}\ln |\mathscr{I} - (4\pi)^2\mathscr{G}_{AA}^{(M)}\mathscr{G}_{A'A}^{(M)}| - \tfrac{1}{2}\ln |\mathscr{I} - (4\pi)^2\mathscr{G}_{BB}^{(M)}\mathscr{G}_{B'B}^{(M)}|$$

$$= \ln\left\{\frac{|\mathscr{I} - (4\pi)^2\mathscr{G}_{AB}^{(M)}\mathscr{G}_{BA}^{(M)}|}{(|\mathscr{I} - (4\pi)^2\mathscr{G}_{AA}^{(M)}\mathscr{G}_{A'A}^{(M)}||\mathscr{I} - (4\pi)^2\mathscr{G}_{BB}^{(M)}\mathscr{G}_{B'B}^{(M)}|)^{1/2}}\right\}.$$

The complete expression for the surface energy of A and B is

$$E_s = -\frac{\hbar}{4\pi i}\oint d\omega \ln\left\{\frac{|\mathscr{I} - (4\pi)^2\mathscr{G}_{AB}^{(M)}\mathscr{G}_{BA}^{(M)}|}{(|\mathscr{I} - (4\pi)^2\mathscr{G}_{AA}^{(M)}\mathscr{G}_{A'A}^{(M)}||\mathscr{I} - (4\pi)^2\mathscr{G}_{BB}^{(M)}\mathscr{G}_{B'B}^{(M)}|)^{1/2}}\right\}. \quad (4.104)$$

This will evidently vanish if $A \equiv B$ and $A' \equiv B'$ as it should, because the surface then disappears. The numerator term in the integrand on the right hand side of eqn. (4.104) gives the Lifshitz interaction energy, as we have seen. We thus have the important result that the Lifshitz interaction energy between two bodies is the difference between the surface energy of the system and the sum of the surface energies of the two bodies taken separately.

This exact result is valid only when the surface energies are evaluated exactly. If we evaluate them to leading order in $\mathscr{G}^{(M)}$, as in §4.6, we can expect some discrepancy in so far as terms of higher order in polarizability are concerned.

4.9 Remarks on approximations and assumptions

The main problem which has concerned us in this chapter might be rephrased as follows: How can one construct a theory of interactions between macroscopic bodies which retains the virtues of the continuum theory of Lifshitz (i.e., the inclusion of many-body aspects through its reliance on measurable macroscopic properties), and which also holds at all distances of separation? As developed here, the solution to this problem is based on taking into account the effect of the finite size of the constituent molecules on the dielectric response of the system. The finite size leads to the useful concept of the dispersion self-energy* of a molecule, and removes the divergence difficulties of Lifshitz's theory at small distances.

The principal assumptions made are worth recapitulating, and are:

(i) The finite size of the molecule is taken into account by assuming a Gaussian distribution of polarizability. This is not a serious limitation. The Gaussian form has been chosen for mathematical convenience only. Any other peaked distribution which models the actual molecular polarizability as defined by eqn. (4.7) will give essentially the same kind of dependence of self-energy and interaction energy on molecular size, with differences in numerical factors only.

(ii) In calculating $\mathscr{G}^{(M)}(\mathbf{r}, \mathbf{r}')$ we have used the solution of Maxwell's equations with infinitely sharp boundaries. There are two separate aspects to this assumption. The first is the question of whether $\mathscr{G}^{(M)} = [\mathscr{I} + 4\pi\mathscr{G}_N]^{-1}\mathscr{G}_N$ is the same as the $\mathscr{G}^{(M)}$ obtained by solving Maxwell's equations. Although a rigorous proof of the identity of the two macroscopic Green functions can not be given, there are strong plausibility arguments (Renne, 1971) for assuming that the identification holds to a high degree of approximation. In dilute systems with point molecules $\mathscr{G}^{(M)}$ can actually be obtained by summing the infinite series in powers of $\mathscr{G}_N$ and the resulting form (Langbein, 1971, 1974) coincides with that obtained

---

* Dispersion self-energy of a molecule is the dispersion analogue of the Born electrostatic self-energy of an ion immersed in a dielectric medium, which has played such a useful role in physical chemistry.

by solving Maxwell's equations. The way molecular size is incorporated into $\mathscr{G}^{(M)}$ as obtained from Maxwell's equations is essentially equivalent to taking the pre-factor $[\mathscr{I}+4\pi\mathscr{G}_N]^{-1}$ in eqns. (4.85) as for point molecules, and the post-factor $\mathscr{G}_N$ for molecules of finite size. This cannot introduce any serious error, since the pre-factor does take into account the screening effect of the molecules which does not depend significantly on molecular size, except at small distances, and at small distances the post-factor takes over.

The second aspect is the assumption of the infinite sharpness of the interfaces. This is clearly unphysical, particularly when one considers distances of separation of the order of molecular size. This assumption renders the results for small separations of qualitative significance only in general, since the exact answer must come from a Green function $\mathscr{G}^{(M)}$ that correctly takes into account the density profiles at the interfaces. Although this is possible in principle (Barber and Perram, 1974), we have not attempted this here.

(iii) Wherever possible the connection between the molecular polarizability and the bulk dielectric constant is taken to be of the form given in eqn. (4.92). This is rigorously valid only for dilute systems, and may give the impression that this kind of analysis is applicable only to dilute systems. However the formulae for the interaction energy of macroscopic bodies are more general than may appear from their manner of derivation.* This is because of the fact that the change of the modes of the electromagnetic field in the presence of macroscopic bodies involve the same secular equations as are obtained here from the microscopic point of view, the only difference being that the components of the Green functions have continuous indices. The latter feature has already been used in replacing sums by integrals. We remark here that eqn. (4.92), or equivalently the Clausius-Mossotti formula, is valid for a surprisingly large range of densities. The paper of Lord Rayleigh (1892) on the bulk dielectric properties of a packed assembly of dielectric spheres and of cylinders had already justified the use of the Lorentz–Lorenz relation (4.92) for connecting microscopic and macroscopic dielectric properties of media as long ago as 1892.

The model developed in §§4.6–4.8 does clarify the role of dispersion forces in interaction and surface energies. To see this we examine the work of Israelachvili (1973) on estimates of the surface energies of liquid hydrocarbons. He made the apparently extreme assumption that Lifshitz's theory is valid right up to distances of separation comparable with molecular sizes, and was able to compute surface energies (i.e., half the magnitude of the Lifshitz interaction energy at a distance of the order of

---

* See Appendix C for a detailed discussion.

molecular size), in extraordinary agreement with experimental data. Even though an element of curve-fitting is required to fix the minimum molecular distance, once determined the same parameter predicts surface energies for a wide range of liquids and the agreement with experiment cannot be regarded as fortuitous. The analysis of §4.8 gives a sound physical basis to Israelachvili's conjecture. Notwithstanding these remarks, it should be borne in mind that the theory developed so far holds only for media whose interactions are dominated by dispersion forces. If permanent dipole interactions are involved as in polar fluids, one can expect some change of orientation of the surface dipoles as two bodies come together.

The relationship between interaction and surface energies as given above is similar in spirit to what had been anticipated by Young (1805), and further expounded by Maxwell (1875) just a century ago.

An analysis of the interaction between two dielectric slabs somewhat akin to the analysis given here has been made by Craig (1973). The important differences between his analysis and that given above is that here the microscopic aspects of the problem, such as molecular size, are taken into account explicitly. Further, his result for the interaction energy involves an integration over a coupling constant (connecting the molecular polarization and the field strength) which occurs in the dielectric constant in a complicated form. No such problem arises in the formalism developed here, since the results are given in terms of the macroscopic Green functions. In fact the integration over coupling constant in Craig's formalism is equivalent to using (4.53).

4.10 Self-energy for the two slab problem

The self-energy of a molecule for the situation depicted in Fig. 2.1 will be useful in our investigation of adsorption phenomena. We give here the main results, and refer to Appendix A for details. Recall that from eqn. (4.63), to lowest order in $\mathscr{G}^{(M)}$ the self-energy of a molecule of polarizability α, size a and position R_j is given by

$$E(j) = -\frac{\hbar}{i} \oint d\omega \, \mathrm{Tr} \, \mathscr{G}^{(M)}(\mathbf{R}_j, \mathbf{R}_j; \omega). \qquad (4.105)$$

The Green function $\mathscr{G}^{(M)}$ is defined by eqns. (4.64) and (4.65) with the boundary conditions that $G^{(M)}$ and $\varepsilon(\partial G^{(M)}/\partial z)$ be continuous across the boundaries. Note that $G^{(M)}$ is the Green function for two interfaces, and not that for a single interface which appears in the analysis of §4.7 and §4.8. The trace of the required dyadic Green function is obtained after some tedious algebra in the form

$$
\operatorname{Tr} \mathscr{G}^{(M)}(\mathbf{R}, \mathbf{R}) = \frac{\alpha(\omega)}{2\pi a^3} \left\{ \frac{1}{\varepsilon_1} \int_{z/a}^{\infty} + \frac{1}{\varepsilon_2} \int_{(z-l)/a}^{z/a} + \frac{1}{\varepsilon_3} \int_{-\infty}^{(z-l)/a} \right\}
$$

$$
\times \operatorname{erfc} |y| \, \mathrm{d}y + \frac{\alpha(\omega)}{4\pi} \int_0^{\infty} \frac{\mathrm{d}\kappa \, e^{-\kappa^2 a^2/4} e^{-z^2/a^2}}{(1 - \Delta_{12}\Delta_{32} e^{-2\kappa l})}
$$

$$
\times \left\{ \frac{\kappa}{\sqrt{\pi} a} \left[\Delta_{12} e^{-\kappa|z|} \operatorname{sgn}(z) + \Delta_{12}\Delta_{32} \, e^{-\kappa(|l-z|+l)} \operatorname{sgn}(l-z) \right] \right.
$$

$$
\times \left[\left(\frac{1}{\varepsilon_1} + \frac{1}{\varepsilon_2} \right) + \left(\frac{1}{\varepsilon_3} - \frac{1}{\varepsilon_2} \right) e^{-l^2/a^2} e^{-l(\kappa - 2z/a^2)} \right]
$$

$$
+ \kappa^2 \left(\left[\frac{1}{\varepsilon_1} e^{(\kappa a/2 + z/a)^2} \operatorname{erfc}(\kappa a/2 + z/a) \right] [\Delta_{12} e^{-\kappa|z|} \theta(-z)
$$

$$
- \Delta_{12}\Delta_{32} e^{-\kappa(|\kappa - z|+l)} \theta(l-z)] + \left[e^{(\kappa a/2 - z/a)^2} \right.
$$

$$
\times \left(\frac{1}{\varepsilon_2} \operatorname{erfc}(\kappa a/2 - z/a) + \left(\frac{1}{\varepsilon_3} - \frac{1}{\varepsilon_2} \right) \operatorname{erfc}(\kappa a/2 - z/a + l/a) \right) \right]
$$

$$
\times \left[-\Delta_{12} \, e^{-\kappa|z|} \theta(z) + \Delta_{12}\Delta_{32} e^{-\kappa(|l-z|+l)} \theta(z-l) \right] \right) \right\}
$$

$$
+ \{ 3 \rightleftharpoons 1, \ z \to l - z \}.
$$

$$
(4.106)
$$

The meaning of $\{3 \rightleftharpoons 1, \ z \to l-z\}$ is that it stands for the whole previous expression in curly brackets with the indices 3 and 1 interchanged and z replaced by $(l-z)$ everywhere in the integral over κ. Using eqn. (4.106) in eqn. (4.105), to leading order in $\mathscr{G}^{(M)}$, we get the self-energy of a molecule as a function of z. From this expression the force field on the molecule can be obtained by differentiation with respect to z.

Although the expression for the self-energy is complicated, limiting forms can be obtained immediately. For a molecule of polarizability α_j in medium 1, with $|z| \gg a_j$ and $l \gg a_j$ where a_j is its size, we get

$$
F(z) = -\frac{\partial E(j)}{\partial z} = -\frac{2\hbar}{\pi} \int_0^{\infty} \frac{\alpha_j(i\xi) \, \mathrm{d}\xi}{\varepsilon_1} \int_0^{\infty} \kappa^3 \, \mathrm{d}\kappa \, \exp(-2\kappa|z|)
$$

$$
\times \left(\frac{\Delta_{12} - \Delta_{32} \exp(-2\kappa l)}{1 - \Delta_{12}\Delta_{32} \exp(-2\kappa l)} \right). \quad (4.107)
$$

When $|z| \gg l$, or $|z| \ll l$, we recover the well-known result for the force between a molecule and a dielectric half-space of dielectric constant ε_2 or

ε_3 depending on which inequality holds. For instance if $|z| \ll l$,

$$F(z) \approx -\frac{3\hbar}{4\pi|z|^4} \int_0^\infty \Delta_{12} \frac{\alpha_1(i\xi)\,d\xi}{\varepsilon_1}. \tag{4.108}$$

If the molecule is immersed in medium 2, the corresponding limiting results are (for $z \gg a_j$, $l \gg a_j$)

$$F(z) \approx -\frac{2\hbar}{\pi} \int_0^\infty \frac{\alpha_j(i\xi)\,d\xi}{\varepsilon_2}$$

$$\times \int_0^\infty \kappa^3 \, d\kappa \left(\frac{\Delta_{12} \exp(-2\kappa z) - \Delta_{32} \exp[-2\kappa(l-z)]}{1 - \Delta_{12}\Delta_{32}\exp(-2\kappa l)} \right). \tag{4.109}$$

These are asymptotic results. The energy of a molecule as also the force on it do not diverge as $z \to 0$ or $z \to l$, but tend to finite values. Note too that we have retained only terms in eqn. (4.105) in $\mathrm{Tr}\,\mathscr{G}^{(M)}$. Higher powers will contribute in general, although these can be expected to be unimportant.

This method can also be extended to give the force on a dielectric sphere immersed in the system, using the effective polarizability of the sphere in place of α_j. This force field can be expected to play a role in consideration of the transport of molecules across membranes. This is in addition to the electrostatic effects which have already been considered by Parsegian (1969).

4.11 A note on the binding energy of a molecular crystal

The binding energy of a molecular crystal has two parts. One part arises from the short range repulsive interactions between the molecules and the other is from the longer range dispersion interactions. The latter can be evaluated from eqn. (4.49). For point molecules this and equivalent expressions for point molecules have been given by Mahan (1965), Lucas (1967), Hüller (1971) and others.

An interesting problem that arises in this connection is the binding energy of rare gas crystals. They crystallize in face centred cubic (FCC) structure. Attempts have been made to find out (Lucas, 1967) whether the energy for FCC structure is less than the hexagonal close-packed (HCP) structure. In both these structures the number of nearest neighbours is the same, so that any difference in energy must arise from the long range dispersion part. For point molecules the dispersion energies turn out to be the same within the limits of computational error.

It has been suggested that structure sensitivity of this energy could arise from inclusion of multipolar interactions rather than the dipole-dipole interactions only. However, if the finite size of the molecules is taken into consideration, as in derivation of eqn. (4.49), a structure sensitivity of the binding energy may be expected to arise.

In the non-retarded limit and with a molecular form factor of the type given in eqn. (4.21), eqn. (4.27) can be written as,

$$\mathcal{G}(\mathbf{r}, \mathbf{r}') = \frac{\alpha(\omega)}{(2\pi)^3} \int d^3 k \, \frac{(\mathbf{k}\,\mathbf{k})}{k^2} e^{i\mathbf{k}\cdot(\mathbf{r}-\mathbf{r}')} e^{-k^2 a^2/4}. \qquad (4.110)$$

A computation using the dyadic Green function of eqn. (4.110) for FCC and HCP structures has been attempted (Mahanty and Richardson, 1975). This is done by writing eqn. (4.49) for N molecules in the form,

$$E_B = -\frac{\hbar}{4\pi i} \oint d\omega \sum_{n=1}^{\infty} \frac{(-1)^{n+1}(4\pi)^n}{n}$$
$$\left(\mathrm{Tr}\, \{\mathcal{G}^{(N)}\}^n - \sum_{j=1}^{N} \mathrm{Tr}\, \{\mathcal{G}(\mathbf{R}_j, \mathbf{R}_j)\}^n \right). \qquad (4.111)$$

From eqn. (4.110) we get

$$\mathrm{Tr}\, \{\mathcal{G}(\mathbf{R}_j, \mathbf{R}_j)\}^n = 3 \left(\frac{\alpha(\omega)}{3\pi^{3/2} a^3} \right)^N. \qquad (4.112)$$

$\mathrm{Tr}\, \{\mathcal{G}^{(N)}\}^n$ can be obtained by using the translational symmetry of the crystal in terms of sums over reciprocal lattice vectors. The details of this calculation may be seen in the paper by Mahanty and Richardson (1975). We give here only the summary of the results.

Equation (4.111) can obviously be written as,

$$E_B = N \sum_{n=2}^{\infty} c_n I_n, \qquad (4.113)$$

where

$$I_n = -\frac{\hbar}{4\pi i\, d^{3n}} \oint [\alpha(\omega)]^n \, d\omega \qquad (4.114)$$

and the coefficient c_n is obtained from the dyadic Green function. Here d is the interatomic distance. If each molecule is treated as an oscillator of frequency ω_0, with $\alpha(\omega) = \alpha(0)/(1 - \omega^2/\omega_0^2)$

$$I_n = (\tfrac{1}{2}\hbar\omega_0) \frac{1}{2\sqrt{\pi}\, d^{3n}} [\alpha(0)]^n \frac{\Gamma\left(\dfrac{2n-1}{2}\right)}{\Gamma(n)}. \qquad (4.115)$$

Table 4.1

		$n = 2$	3	4	5	6	7
$I_n/(\tfrac{1}{2}\hbar\omega_0)$		$3{\cdot}725 \times 10^{-5}$	$3{\cdot}412 \times 10^{-7}$	$3{\cdot}461 \times 10^{-9}$	$3{\cdot}691 \times 10^{-11}$	$4{\cdot}061 \times 10^{-13}$	$4{\cdot}534 \times 10^{-15}$
$c_n I_n \times 10^3/(\tfrac{1}{2}\hbar\omega_0)$	FCC	$-1{\cdot}597$	$0{\cdot}537$	$-0{\cdot}153$	$0{\cdot}0419$	$-0{\cdot}0133$	$0{\cdot}00307$
	HCP	$-1{\cdot}537$	$0{\cdot}517$	$-0{\cdot}148$	$0{\cdot}0404$	$-0{\cdot}0109$	$0{\cdot}00294$
Lucas (1967)		$-0{\cdot}625$	$0{\cdot}0395$	$-0{\cdot}0041$	—	—	—

Table 4.1 gives for FCC and HCP the values of I_n and $c_n I_n$ for $n = 2, 3, 4,$ ..., 7, with the oscillator model parameter $\alpha(0)$ and d chosen for solid neon, and $(a/d)^2 = 0{\cdot}125$. The earlier calculation of Lucas (1967), also given in the table, does not differentiate between the two structures. The difference between the energies of the two structures, which is non-existent in the point molecule model, arises due to the spread of molecular polarizability in this formalism. The same problem has been resolved from a perturbation theoretic approach by Niebel and Venables (1974). An application of the idea of dispersion self-energy of an impurity in a polar crystal and to some aspects of the polaron problem has been made by Mahanty and Paranjape (1976).

References

Barber, M. N. and Perram, J. W. (1974). *Mol. Phys.* **28,** 131.
Beattie, A. M., Stoddart, J. C. and March, N. H. (1971). *Proc. Roy. Soc. A* **326,** 97.
Bethe, H. A. (1947). *Phys. Rev.* **72,** 339.
Craig, R. A. (1972). *Phys. Rev. B* **6,** 1134.
Craig, R. A. (1973). *J. Chem. Phys.* **58,** 2988.
Fowkes, F. M. (1964). *Ind. Eng. Chem.* **12,** 40.
Fowkes, F. M. (1968). *J. Coll. Interface Sci.* **28,** 493.
Hüller, A. (1971). *Z. Physik* **241,** 340.
Israelachvili, J. N. (1972). *Proc. Roy. Soc. A* **331,** 39.
Israelachvili, J. N. (1972). *J. Chem. Soc., Faraday II* **69,** 1729.
Jehle, H. (1965). *Advan. Quantum Chem.* **2,** 195.
Kohn, W. and Sham, L. J. (1965). *Phys. Rev. A* **140,** 1133.
Kubo, R. (1957). *J. Phys. Soc. Japan* **12,** 570.
Langbein, D. (1971). *J. Phys. Chem. Solids* **32,** 133.
Langbein, D. (1974). "Theory of van der Waals Attraction," Springer Tracts in Modern Physics. Springer, Berlin. **72,** 139 pp.
Lundqvist, S. and Sjölander, A. (1964). *Ark. Fys.* **26,** 17.
Linder, B. (1964). *J. Chem. Phys.* **40,** 2003.
Linder, B. and Rabenold, D. A. (1972). *Advan. Quantum Chem.* **6,** 203.
London, F. (1942). *J. Chem. Phys.* **46,** 305.
Lucas, A. A. (1967). *Physica* **35,** 353.
Mahan, G. D. (1965). *J. Chem. Phys.* **43,** 1569.
Mahanty, J. (1974). *Il Nuovo Cimento B* **22,** 110.
Mahanty, J. and Ninham, B. W. (1973). *J. Chem. Phys.* **59,** 6157.
Mahanty, J. and Ninham, B. W. (1974). *J. Chem. Soc., Faraday II* **70,** 637.

Mahanty, J. and Richardson, D. D. (1975). *J. Phys. C* **8,** 1322.

Mahanty, J. and Paranjape, V. V. (1976). *Phys. Rev.* B**13,** 1830.

Maxwell, J. C. (1875). "Capillary Action," Encyclopaedia Britannica 9th ed. updated by Rayleigh in 10th ed.

McLachlan, A. D. (1963a). *Proc. Roy. Soc. A* **271,** 387.

McLachlan, A. D. (1963b). *Proc. Roy. Soc. A* **274,** 80.

Mitchell, D. J. and Richmond, P. (1973). *Chem. Phys. Letts.* **21,** 113.

Mitchell, D. J. and Richmond, P. (1974). *J. Coll. Interface Sci.* **46,** 118, 128.

Niebel, K. S. and Venables, J. A. (1974). *Proc. Roy. Soc.* A**336,** 365.

Parsegian, V. A. (1969). *Nature* **221,** 844.

Power, E. A. (1966). *Am. J. Phys.* **34,** 516.

Rayleigh, Lord (1892). *Phil. Mag.* **34,** 481.

Richardson, D. D. (1975). *J. Phys.* A**8,** 1828.

Renne, M. J. (1971). *Physica* **56,** 125.

Welton, T. A. (1948). *Phys. Rev.* **74,** 1157.

Young, T. (1805). *Phil. Trans.* **65,** 1.

Chapter 5

Geometry and Anisotropy

So far our detailed theoretical considerations on continuum theories of interactions have been limited in the main to half-spaces. Here the example of hydrocarbon–water interactions discussed in §3.3 indicates a quite complicated and delicate interplay between material properties as revealed by ultraviolet, infrared and microwave spectra, and of temperature and retardation. According to the Hamaker theory of interactions the effects of geometry and material properties are separable. If we recall that for hydrocarbon–water the large zero frequency contribution to the total force is predominantly an entropically driven force, i.e., has to do with orientation forces, we must expect that geometry and material properties will be linked. This of course was already recognized by Hamaker (1937), who restricted his analysis of condensed media interactions to situations where orientation forces were absent. The advantage of the Lifshitz approach is that it provides a first step towards tackling this extremely difficult problem. Even if temperature-dependent forces are excluded from consideration, pairwise summation as developed by Hamaker leads to erroneous results in general.

5.1 Lamellar systems

Triple films

Consider first the system illustrated in Fig. 5.1. This serves a model for semi-infinite half-spaces of substance 1 coated with a medium 2, as in the

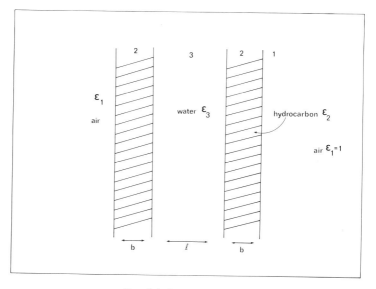

FIG. 5.1. Model soap film.

experiment of Israelachvili and Tabor (see §3.2). Equally it represents a model for a soap film, where medium 3 is salt water, and media 2 are hydrocarbon layers, medium 1 air, or a crude model for two biological cell membranes "2" in close proximity (Parsegian and Gingell, 1972; Parsegian, 1973). The extension of van Kampen's method to this problem is quite straightforward, involving through an extra boundary condition only a little more algebra than that in §2.4 (Ninham and Parsegian, 1970a). In the limit of zero temperature and neglecting retardation we have for the energy of interaction per unit area

$$E(l, b) = \frac{\hbar}{4\pi^2} \int_0^\infty x \, dx \int_0^\infty \ln\left(1 - \Delta_{31}^2(b)e^{-2lx}\right) d\xi, \qquad (5.1)$$

where

$$\Delta_{31}(b) = \frac{\Delta_{32} + \Delta_{21}e^{-2bx}}{1 + \Delta_{32}\Delta_{21}e^{-2bx}}. \qquad (5.2)$$

For model soap films where medium 1 is air, or where medium 3 is a vacuum, the effects of temperature are usually negligible and can be safely ignored in most circumstances. In the limit $b \to \infty$, (5.1) reduces to the Lifshitz dispersion energy associated with a single layer of medium 3 separating two semi-infinite media 2, i.e.,

$$E(l, b \to \infty) = \frac{\hbar}{4\pi^2} \int_0^\infty x \, dx \int_0^\infty \ln\left(1 - \Delta_{32}^2 e^{-2lx}\right) d\xi. \qquad (5.3)$$

Similarly when $b \to 0$, eqn. (5.1) reduces to the energy of a single layer "3" separating two semi-infinite media "1." This is because the zero from which energies are measured is chosen to be a situation in which two layers of fixed thickness b are separated by an infinite distance. For soap films this corresponds to a situation where the thickness b of hydrocarbon is assumed not to vary appreciably with separation. The dispersion force is obtained by partial differentiation of (5.1) with respect to l.

We now indicate in what sense eqn. (5.1) generalizes the former methods (involving pairwise summation of London forces) for calculation of Hamaker constants. To see this consider the single layer result (5.3). Since $\Delta_{32}^2 < 1$ and the major contribution to the x integral comes from the region $x \approx 0$, we can write as before

$$\ln (1 - \Delta_{32}^2 e^{-2lx}) \approx -\Delta_{32}^2 e^{-2lx} \tag{5.4}$$

and carry out the integration with respect to x to obtain

$$E(l, b \to \infty) = E_{23}(l) \approx -\frac{\hbar}{16\pi^2 l^2} \int_0^\infty \left(\frac{\varepsilon_3 - \varepsilon_2}{\varepsilon_3 + \varepsilon_2}\right)^2 \mathrm{d}\xi. \tag{5.5}$$

Pairwise summation gives for the dispersion energy per unit area for medium "3" of thickness l separated by semi-infinite media 2 the result (§1).

$$E_{23}(l) \approx -\frac{\hbar\bar\omega}{16\pi^2 l^2} \equiv -\frac{A}{12\pi l^2}, \tag{5.6}$$

where $\bar\omega$ is the characteristic frequency of the two media. The form of the more general triple layer result can now be understood. Again expanding the logarithm and retaining only leading terms we have

$$\ln (1 - \Delta_{31}^2(b)e^{-2lx}) \approx -(\Delta_{32}^2 e^{-2lx} + \Delta_{21}^2 e^{-(2l+4b)x} + 2\Delta_{32}\Delta_{21}e^{-2(l+b)x}). \tag{5.7}$$

Substituting this expression in (5.1) we have for the energy per unit area of the triple film the approximate expression

$$E(l, b) \approx -\frac{\hbar}{16\pi^2} \int_0^\infty \left(\frac{\Delta_{32}^2}{l^2} + \frac{\Delta_{21}^2}{(l+2b)^2} - \frac{2\Delta_{23}\Delta_{21}}{(l+b)^2}\right) \mathrm{d}\xi. \tag{5.8}$$

By comparison of (5.5) and (5.6) it can be seen that to leading order this is what one might have expected from a consideration of two-body energy summation, i.e., the dispersion energy can be regarded approximately as (1) the energy of interaction across a layer "3" of thickness l separating two semi-infinite media 2, plus (2) the energy of interaction across a layer "2" of thickness $(l+2b)$ which separates two semi-infinite media 1, minus (3) the energy of interaction across a layer 2 of thickness $l+b$ which separates two media 1 and 3 taken twice.

Soap films

Numerical calculations for soap films have been made by Ninham and Parsegian (1970a). Several features deserve note. In the past it had been usual to analyse film experiments in terms of some expression like (see eqn. (1.28))

$$E = -\frac{A_{\text{eff}}}{12\pi l^2}\left(1 - \frac{l^2}{(l+b)^2} + \frac{l^2}{(l+2b)^2}\right), \tag{5.9}$$

where A_{eff} is constant with film thickness (neglecting retardation). Comparison with (5.8) shows that as the film thins A_{eff} is not constant, but should vary with separation l. As $l \to 0$, A_{eff} tends towards the value for hydrocarbon–water interactions, and as $l \to \infty$ towards the value for water vacuum interactions. As already noted this was in fact observed directly by Israelachvili and Tabor (1972). In soap film studies which serve as an experimental model system for studying the validity of DLVO colloid stability theory,* "anomalous behaviour" is observed at soap film thicknesses between 40 and 50 Å (Huisman and Mysels, 1964) and again at large thicknesses (Lyklema and Mysels, 1970). Theoretically, given assumptions about the thickness of the soap molecule chains, one does predict from (5.1) some "anomaly" in van der Waals forces at these distances (Ninham and Parsegian, 1970a). Given that the double-layer calculations which give the repulsive force required to balance van der Waals attractive forces are correct, the observed effect goes in the *opposite* direction to that predicted, and in any case is far too small. The soap film experiments have remained ambiguous. However, their interpretation has depended critically upon the correct assignment of the thickness of the hydrocarbon layers, the Debye–Hückel screening parameter, and other factors. For further references to soap film work see Napper and Hunter (1975).

Another theoretical prediction deserves note. In thin film experiments the film water is in equilibrium with up to twice molar salt concentration (NaCl, LiCl). The electrostatic potential across the water layer due to the hydrophobic ends of the soap molecules makes the total ion concentration in the film higher than that in the bulk reference solution. Because of the inequality of cation and anion concentration in this layer only a reasonable guess is possible for n_{w}, the refractive index of the water. One can ask what effect cation concentration will have on dispersion forces.

---

* DLVO refers to the theory of Deryaguin and Landau (1941) and Verwey and Overbeek (1948).

The effect of a change in salt concentration is to increase the refractive index of water from $n_w^2 = 1\cdot78$ to $1\cdot9$ for 5M NaCl. This represents only a change of 3% in refractive index. For a thin film of hydrocarbon-water-hydrocarbon the corresponding change in dispersion energy is up to 30%. The index of refraction is related to the number of electrons per ion, while electrostatic behaviour is determined by the net charge. One can therefore expect that changes in dispersion energy with salt concentration can be different for ions of like charge but different size, e.g., Li^+ and Na^+. This effect is probably not a major determinant in observed ion-specific increases in attractive energy.

Hydrocarbon films in water

The essential qualitatively different features and complexity of interactions between organic materials in water as opposed to interactions in vacuo or with inorganic materials can be illustrated by an analysis of two films of hydrocarbon in water (Parsegian and Ninham, 1971). For this model system the effects of temperature are very important because of the similarity in the uv spectra of the interacting media, while for soap films—where a vacuum is involved—the temperature-dependent forces are relatively small. In Fig. 5.1 we take media 1 and 3 to be pure water and media 2 to be hydrocarbon layers of thickness 50 Å. The generalization of (5.1) to include temperature and retardation effects is straightforward (Ninham and Parsegian, 1970a; Parsegian and Ninham, 1971), and is given below. For the present discussion we require only that this generalization can be written in a form

$$F(l, b) \equiv -\frac{A(l, b)}{12\pi l^2}. \qquad (5.10)$$

In the limit $T \rightarrow 0$, and in the absence of retardation the full expression reduces to (5.1). The free energy due to van der Waals interaction between two model hydrocarbon layers is plotted as a function of separation l in Fig. 5.2a (Parsegian and Ninham, 1971). The two curves correspond to the exact result (5.10) or the approximation (5.1) which neglects temperature. In the range of separation $200\,\text{Å} < l < 1000\,\text{Å}$ the approximations based on neglect of temperature underestimate the energy by a factor of 5 to 200. The calculation is carried out by the full formula (5.10), bottom line. Essentially the same curve results when retardation is neglected. The top line is a calculation in zero temperature approximation (5.1). Essentially the same curve results from the pairwise summation form (5.9). Note the logarithmic scale of the ordinate and hence order of magnitude error in estimates due to neglect of temperature. Model dielectric data for water and hydrocarbon are taken from

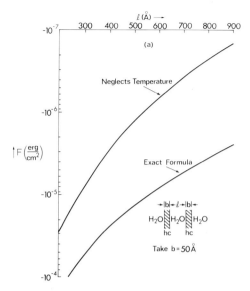

FIG. 5.2. (a) Free energy of van der Waals interaction between two planar hydrocarbon layers in water vs. separation between the layers. [After Parsegian and Ninham (1971).]

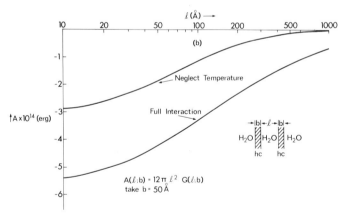

FIG. 5.2. (b) Free energy coefficients $A(l, b)$ vs. spacing l. [From Parsegian and Ninham (1971).]

Parsegian and Ninham (1971). Recent revised infrared data (Parsegian and Gingell, 1972) will not affect computed curves significantly. Similarly in Fig. 5.2b we plot the function $A(l, b, T) = 12\pi l^2 F(l, b, T)$ defined by (5.10) in order to condense the wide range of energies between $10\,\text{Å} < l < 500\,\text{Å}$. The neglect of temperature effects gives an estimate less than or about half the complete expression. The functional form of the $T = 0$

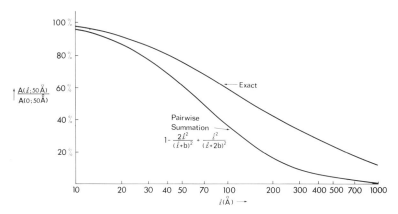

FIG. 5.3. Comparison of the result of pairwise summation with the more general form. [From Parsegian and Ninham (1971).]

curve is dramatically different from the exact result. In the hypothetical limit $l \to 0$, we have $A(0; b) \approx 5 \times 10^{-14}$ erg in agreement with Haydon's measured line. At all separations the neglect of the effect of temperature is a serious omission. We note also (Fig. 5.3) that the decay of $A(l, b)$ vs. $A(0, b)$ is rather different from that predicted by a pairwise summation (5.9). Even if pairwise summation predicted the right Hamaker constant at $l = 0$, it would still predict an incorrect dependence of the van der Waals energy on separation l. The interaction is of much longer range than would be expected from addition of r^{-6} interactions at ultraviolet frequencies. The algebraic form of pairwise summation differs clearly from the more general form. The top line shows relative decay of the coefficient $A(l, 50\,\text{Å})/A(0, 50\,\text{Å})$ vs. spacing l from (5.9). The bottom line shows the relative decrease expected from pairwise summation. Note the difference in the shape of the curves as well as the underestimate from r^{-6} summations at large separations, even though the lower line neglects retardation damping.

It is of interest too to examine in detail the breakdown of additivity, the nature of retardation and effects of temperature. These are illustrated in Fig. 5.4. Underlying the total value of $A(l, b)$ (solid line) are shown the contributions from (i) zero or microwave relaxation frequencies ($n = 0$, $\xi_n = 0$ term); (ii) from infrared frequency fluctuations ($2 \cdot 4 \times 10^{14} \le \xi_n < 3 \cdot 8 \times 10^{15}$ rad/sec); and (iii) from the ultraviolet region taken as $\xi_n > 3 \cdot 8 \times 10^{15}$ rad/sec. We consider each range of frequencies in turn.

(1) The low frequency ($\xi_n = 0$) contribution is always the most important of the three and dominates completely for large separations. The temperature-dependent low-frequency contribution is most important.

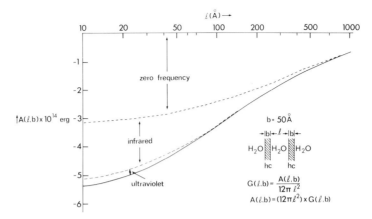

FIG. 5.4. Decomposition of the free energy into different frequency regimes. [From Parsegian and Ninham (1971).]

The ultraviolet contribution is negligible after $l = 100$ Å due to retardation. The zero frequency contribution ($\xi_n = 0$) when isolated from the sum over all frequencies which constitute the whole interaction is (see below),

$$A_{n=0}(l, b) = -(3k_B T/4) \int_0^\infty x \, dx \, \ln\left(1 - \Delta_{ww}^2(\text{eff})e^{-x}\right), \tag{5.11}$$

where

$$\Delta_{ww}(\text{eff}) = \Delta_{whc} \frac{(1 - e^{-xb/l})}{(1 - \Delta_{whc}e^{-xb/l})} \tag{5.12}$$

and

$$\Delta_{whc} = \frac{\varepsilon_w(0) - \varepsilon_{hc}(0)}{\varepsilon_w(0) + \varepsilon_{hc}(0)} \tag{5.13}$$

and susceptibilities are taken at zero frequency. From (5.7) and (5.9) a necessary condition that the form of interaction given by pairwise summation hold is that $\Delta \ll 1$. Here for water and hydrocarbon we have

$$\varepsilon_w \approx 80 \gg \varepsilon_{hc} \approx 2, \qquad \Delta_{whc} \approx 1 \tag{5.14}$$

and in the limit $\Delta_{whc} \to 1$, (5.12) gives $\Delta_{ww}(\text{eff}) = \Delta_{whc}$. Hence the free energy contribution $F_{n=0}$ is essentially independent of slab thickness b, and separation. There is a weak dependence of $A_{n=0}(l, b)$ on b/l which can be observed in Fig. 5.4.

(2) At frequencies ξ_n characteristic of infrared frequencies the ratios Δ are much less than unity and conditions of negligible retardation hold for the range of separations considered. The decay here is like that obtained by the method of pairwise summation.

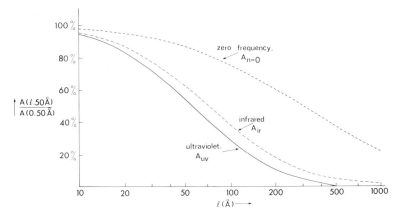

FIG. 5.5. Relative decay of the contributions of different frequencies. [From Parsegian and Ninham (1971).]

(3) The ultraviolet fluctuations decay fastest. Data used for plotting curves in Figs. 5.3 and 5.4 are incorrect (Parsegian and Gingell, 1972) and underestimate the actual contribution from the ultraviolet and visible regions, but the qualitative features remain.

The relative decrease of each frequency band with spacing is compared in Fig. 5.5. The high frequency terms drops to negligible values at $l \gtrsim 200$ Å, whereas the zero frequency term drops off very slowly. Contributions from different frequencies decay at different rates. Ultraviolet contributions decay fastest because they are subject to severe geometric and retardation constraints. The infrared, subject to geometric constraints, primarily decays like a pairwise additive interaction. The dominant low-frequency term is strictly non-additive and behaves differently to the high frequencies. The most curious feature of the triple layer hydrocarbon-water system is that the non-additivity of the van der Waals forces here makes the hydrocarbon layers appear much thicker than would otherwise be expected, and the attraction between two thin membranes in pure water resembles that between two semi-infinite planes of hydrocarbon across water. In the presence of dissolved salt, this conclusion is dramatically altered, as will be seen in Chapter 7.

5.2 Other many-layered structures

A catalogue of formulae for other multiple-layered structures has been developed by Parsegian and Ninham (1972). Apart from satisfying a consuming passion for elementary algebra the main motive for the

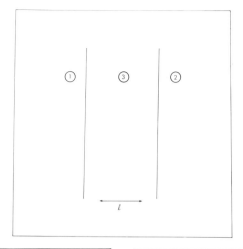

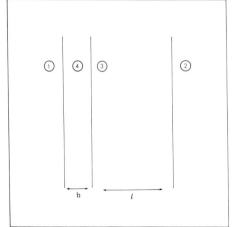
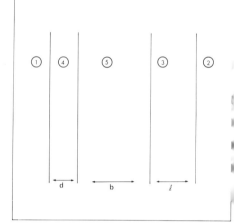

FIG. 5.6. More complicated planar geometries. l varies in each case but other distances are fixed.

construction of such a catalogue is that by application of such formulae one may hope to develop an understanding of the attractive forces between membrane structures like the interaction of biological cells, whose thin lipid-like membranes are usually coated with adsorbed layers of protein and saccharides in water with an extracellular medium outside. Studies of the consequences of the formulae given here have been made by Parsegian and Gingell (1972) and Parsegian (1973) who examined the adhesion of model cell surfaces to substrata-like glass. These studies are deficient in so far as they ignore the changes of surface energies on approach of the two model surfaces, the change in thickness of adsorbed layers with close proximity, and the important effect of the influence of ionic conduction processes which act to screen temperature-dependent forces. But extension of principles developed by Parsegian and Gingell should provide some insights into the nature of cellular adhesion. The main conclusion that emerges from quite an extensive series of calculations is that a balance of van der Waals and electrostatic forces give an energy minimum sufficient to account for adhesion.

The general formulae can easily be summarized. Recall first that for the situation in Fig. 5.6a, the general result of Dzyaloshinskii, Lifshitz and Pitaevskii (1961) can be written as

$$F(l, T) = \frac{k_B T}{8 \pi l^2} \sum_{n=0}^{\infty}{}' \int_r^\infty x \, dx \{\ln (1 - \Delta_{31}^R \Delta_{32}^R e^{-x}) + \ln (1 - \bar{\Delta}_{31}^R \bar{\Delta}_{32}^R e^{-x})\}, \qquad (5.15)$$

where

$$\Delta_{jk}^R = \frac{s_k \varepsilon_j - s_j \varepsilon_k}{s_k \varepsilon_j + s_j \varepsilon_k}, \qquad \bar{\Delta}_{jk}^R = \frac{s_k - s_j}{s_k + s_j}, \qquad s_k = \sqrt{p^2 - 1 + \varepsilon_k/\varepsilon_3}$$

$$r = 2l\sqrt{\varepsilon_3} \xi_n/c, \qquad \xi_n = 2\pi n k_B T/\hbar, \qquad \varepsilon = \varepsilon(i\xi_n). \qquad (5.16)$$

For the situation in Fig. 5.6b, the generalization for fixed l is simply to replace $\bar{\Delta}_{31}^R, \Delta_{31}^R$ by

$$\Delta_{31}^R \to \Delta_{31}^R(b) \equiv \frac{\Delta_{34}^R + \Delta_{41}^R e^{-(xbs_4/pl)}}{1 + \Delta_{34}^R \Delta_{41}^R e^{-(xbs_4/pl)}}. \qquad (5.17)$$

Similarly for Fig. 5.6c the corresponding generalization is $\Delta_{31}^R \to \Delta_{31}^R(l, b)$, where

$$\Delta_{31}^R(l, b) = \frac{\Delta_{35}^R + \Delta_{51}^R(b)e^{-(x \, ds_5/pl)}}{1 + \Delta_{35}^R \Delta_{51}^R(b)e^{-(x \, ds_5/pl)}} \qquad (5.18)$$

and $\Delta_{51}(b)$ is defined by (5.17). By successive application of (5.17) and (5.18) one can write down results for any number of layers. Only in the combined limits that all coating thicknesses go to zero, $T \to 0$, all $\varepsilon_j \approx \varepsilon_3$ and when retardation is neglected do the results go over to the form obtained by pairwise summation.

5.3 Infinite multilayers

While formulae like those above may have some application to the study of cell adhesion, infinite multilayers usually represent a cleaner experimental system whose potentiality for studying molecular forces has

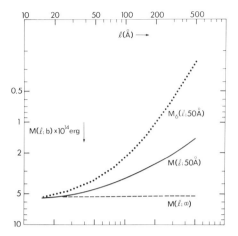

FIG. 5.7. Free energy coefficient $M(l, b)$ vs. spacing l for a multilayer. [From Ninham and Parsegian (1970b).]

never been exploited in full. Multiple layers occur in various clays whose study is (not unnaturally) of considerable interest in soil science and clay chemistry. In biology they have a certain universality, occurring in and of interest in the cell cytoplasm, in membraneous elements of mitochondria, nuclei, chromosomes, endoplasmic reticula, Golgi apparatus, chloroplasts and nerve myelin. Alternating planes of soap and water occur in the lamellar phase of certain liquid crystals which have been studied by Parsegian (1969). In such systems the free energy of formation of the system is of central interest. The general formula for an infinite multilayer has been given by Ninham and Parsegian (1970b) and by Langbein (1972a). While not impossibly messy, it is so considerably lacking in aesthetic appeal that we prefer to refer to the original papers for details. The results of calculations carried out for a hydrocarbon–water multiple layer structure is indicated in Fig. 5.7. The free energy of formation per layer of hydrocarbon is written in the form

$$F(l, b) = -\frac{M(l, b)}{12\pi l^2} \tag{5.19}$$

which can be compared with eqn. (1.34) obtained from pairwise summation. The calculation is for hydrocarbon layers of thickness b separated by water layers. The approximation $M_6(l, 50 \text{ Å})$ is the form expected from assuming pairwise additivity of (r^{-6}) interactions (eqn. 1.34). The dependence of $M_6(l, 50 \text{ Å})$ is clearly different from the correct form $M(l, 50 \text{ Å})$ due to the failure of additivity of van der Waals interactions. For comparison the limiting case for $b \to \infty$ is shown illustrating negligible retardation effects.

Notice again from Fig. 5.7 that neglect of temperature-dependent terms gives a great underestimate of the full interaction free energy. The neglect of retardation in summing r^{-6} interactions should be a source of overestimate. The resulting interaction is of much longer range than that expected for a stack of thin layers. Indeed these layers may be better approximated by infinitely thick layers at large separation. This is a peculiar property of two materials of vastly different static dielectric constants, a circumstance which frequently obtains in biological situations. The same gross features will obtain for clay plates $[\varepsilon(0) \approx 5]$ in pure water $[\varepsilon(0) \approx 80]$.

An amusing consequence of the very long range interaction for a multilayer in water has been pointed out by Perram and Smith (1973). Like the soap film experiments, certain anomalies observed in clay swelling have so far resisted valiant theoretical attempts on the part of many competent people. It seems highly probable that the proper explanation of the observed behaviour of soap films rests with the inner nature of the electrostatic double layer. Since the double layer lies outside our commission, we drop the matter forthwith.

The sometimes apparent failure to predict particle spacing of a simple picture based on a balance between electrostatic and van der Waals forces for plate-like (clay), rod-shaped (tobacco mosaic virus, muscle) and spherical (e.g. latex spheres) is not mysterious, and has to do with the neglect of statistical mechanical aspects (two phase equilibrium) of these problems and occasionally to the neglect of hydration effects. This had been pointed out by Langmuir (1938) and by Onsager (1941) a very long time ago. See e.g. Forsyth, Marčelja, Mitchell and Ninham (1977), Parsegian and Brenner (1976) and Marčelja, Mitchell and Ninham (1976).

5.4 Spheres

Since the world does not consist entirely of planes or points, we turn now to an examination of the effects of other geometries on van der Waals forces. Evidently this is a key problem—the precise magnitude of van der Waals forces as a function of separation of interacting bodies is relevant to considerations involving self-assembly in biology. As mentioned in §1.8, Langbein was the first to improve on Hamaker theory to study the influence of geometry on dispersion forces, and developed the two distinct approaches (Langbein, 1969, 1970, 1971a,b) mentioned earlier. We consider first spheres. For two spheres, radii b_1, b_2, susceptibilities $\varepsilon_1, \varepsilon_2$ separated by a vacuum a distance R apart, where R is the centre-to-centre separation, the exact non-retarded result (Langbein,

1971a) has already been given in eqn. (1.96). If the intervening medium has susceptibility ε_3, the sole change is that $\eta_i(m)$ defined by eqn. (1.94) is modified to the form

$$\eta_i(m) = \frac{m(\varepsilon_i - \varepsilon_3)}{m\varepsilon_i + (m+1)\varepsilon_3}. \tag{5.20}$$

In the form (1.96) the "improvement" on Hamaker theory through inclusion of screening and multiple reflection terms is not immediately apparent. The sums $C_\nu(m, n)$ become increasingly complicated with increasing ν and the series can be expressed as a power series in $(b_1 b_2 / R)$ as in Mitchell and Ninham (1972),

$$E = -\frac{\hbar}{2\pi} \int_0^\infty d\xi \left\{ 24 \left(\frac{b_1 b_2}{R^2}\right)^2 \frac{\Delta_{13}\Delta_{23}}{(3-\Delta_{13})(3-\Delta_{23})} \right.$$
$$\left. + 120 \left[\left(\frac{b_1}{R}\right)^3 \left(\frac{b_2}{R}\right)^5 \frac{\Delta_{13}\Delta_{23}}{(3-\Delta_{13})(5-\Delta_{23})} + (2 \to 1) \right] + 0\left(\frac{1}{R^{12}}\right) \right\}. \tag{5.21}$$

The exact form (1.20) is useful for precise computation only in the large distance limit $R \gg b_1, b_2$. However some useful bounds can be obtained. If the multiple reflection terms $(\nu > 1)$ are excluded (and this is certainly a good approximation for planes), Langbein obtains two bounds

$$|E| \le \frac{\hbar\bar{\omega}}{2\pi} b_1 b_2 \left\{ \frac{1}{2} \left(\frac{1}{[R^2 - (b_1 + b_2)^2]} + \frac{1}{[R^2 - (b_1 - b_2)^2]} \right) \right.$$
$$\left. - \frac{1}{R^2 - b_1^2} - \frac{1}{R^2 - b_2^2} + \frac{1}{R^2} \right\} \tag{5.22}$$

$$|E| \ge \frac{\hbar\bar{\omega}}{4\pi} \left\{ \frac{b_1 b_2}{R^2 - (b_1 + b_2)^2} + \frac{b_1 b_2}{R^2 - (b_1 - b_2)^2} \right.$$
$$\left. + \frac{1}{2} \ln\left(\frac{R^2 - (b_1 + b_2)^2}{R^2 - (b_1 - b_2)^2} \right) \right\}, \tag{5.23}$$

where

$$\hbar\bar{\omega} = \hbar \int_0^\infty \left(\frac{4\pi}{3}\right) A_{123}. \tag{5.24}$$

The lower bound (5.23) is identical with Hamaker's result (1.42). In the case of small separations $l = R - b_1 - b_2 \ll R$, both bounds coincide and we have

$$E \approx -\frac{\hbar\bar{\omega}}{8\pi} \frac{b_1 b_2}{l(b_1 + b_2)}. \tag{5.25}$$

If all terms in ν are kept in (1.96), a further complicated bound can be obtained which reduces in the limit of $R \to (b_1 + b_2)$, $l \to 0$, to

$$|E| \le \frac{\hbar}{8\pi} \int_0^\infty d\xi \sum_{\nu=1}^\infty \frac{1}{\nu^3} (\Delta_{13} \Delta_{23})^\nu \frac{b_1 b_2}{l(b_1 + b_2)}. \qquad (5.26)$$

That is, in the limit $l \to 0$ to Lifshitz expression for planes modified by a geometric factor. This, although not proved, appears to be a quite general result for arbitrary shapes; i.e., in the limit of very close separation the dispersion energy reduces to the planar result times a geometric factor obtainable from Hamaker's method. The same conclusion appears to hold even when effects of temperature are included.

The bounds (5.22), (5.23) and (5.26) are sufficient to provide a reasonable description of interaction energies for systems dominated by dispersion forces alone. For organic materials in water where $\Delta_{13} \sim 1$, they are not very precise. The transition from a $1/l$ dependence to $1/R^6$ represents a rapid change in order of magnitudes, and it is necessary to go further for applications in colloid science. Mitchell and Ninham (1972) have approached the problem via van Kampen's method and solved the normal mode problem in spherical bipolar coordinates. These are a natural coordinate system for spheres at close separation. Their results are as follows: Let $b_1 = b \operatorname{csch} \zeta$, $b_2 = b \operatorname{csch} \eta$, $R = (b^2 + b_1^2)^{1/2} + (b^2 + b_2^2)^{1/2}$, so that for small separations

$$\zeta = \frac{(\bar{b}l)^{1/2}}{b_1}, \qquad \eta = \frac{(\bar{b}l)^{1/2}}{b_2}, \qquad (\zeta + \eta) \sim \left(\frac{2l}{\bar{b}}\right)^{1/2}, \qquad \bar{b} = \frac{2b_1 b_2}{(b_1 + b_2)}. \qquad (5.27)$$

Then as $l/\bar{b} \to 0$, we have

$$E \sim \frac{\hbar}{4\pi(\zeta + \eta)^2} \int_0^\infty d\xi \int_0^\infty x \, dx \ln[1 - \Delta_{13} \Delta_{23} e^{-x}]$$
$$+ \frac{\hbar}{4\pi} \frac{\zeta \ln \zeta}{(\zeta + \eta)} \int_0^\infty [1 - \Delta_{13}] \ln[1 - \Delta_{13} \Delta_{23}] \, d\xi$$
$$+ \frac{\hbar}{4\pi} \frac{\eta \ln \eta}{(\zeta + \eta)} \int_0^\infty (1 \to 2) \, d\xi + \cdots. \qquad (5.28)$$

Provided $\Delta_{ij} \ll 1$, the logarithms can be expanded in powers of Δ, and using (5.27) we recover

$$E \sim -\frac{\hbar}{16\pi} \left\{ \frac{\bar{b}}{l} + 2 \ln\left(\frac{l}{\bar{b}}\right) \right\} + \cdots \int_0^\infty \Delta_{13} \Delta_{23} \, d\xi \qquad (5.29)$$

which agrees with Hamaker's result to terms of leading order (eqn. 1.43). At large separations they recover (5.21).

We make two remarks concerning (5.28). First, the leading term, which is the non-retarded Lifshitz formula modified by a geometric factor, can be positive or negative, and for two like spheres is always attractive. Second, for any two like spheres the perturbation terms always have the opposite sign to the leading term. This is to be expected, since intuitively one expects the two similar spheres to interact as planes when very close together, and to behave more nearly as point dipoles with increasing separation; i.e., the energy of interaction should diminish more rapidly with separation than the planar approximation predicts. Both features are in fact built into the Hamaker formula. But the continuum result goes further. Consider for example the interaction of two very small drops of oil (ε_1) across water (ε_3), and suppose their radii are equal. If the spheres are close, we have at finite temperature

$$
F(l, T) = -\frac{k_B T}{8\zeta^2} \sum_{n=0}^{\infty}{}' \left\{ \left[\sum_{\nu=1}^{\infty} \left(\frac{\Delta_{13}^2}{\nu^3} \right) \right] - 4(\zeta^2 \ln \zeta)(1 - \Delta_{13}) \right.
$$
$$
\left. \times \ln (1 - \Delta_{13}^2) + \cdots \right\}. \quad (5.30)
$$

In the case of planes for hydrocarbon–water interactions, as we have seen, the zero frequency term contributes more than half the whole free energy. For spheres the situation is more complicated. We decompose the whole free energy into two terms as before and write

$$
F = F_{n=0} + F_{n \neq 0}. \quad (5.31)
$$

For the infrared and ultraviolet correlations ($F_{n \neq 0}$), we have $\Delta_{13} \ll 1$ over the whole frequency range. Consequently we can keep only leading terms in (5.30). This gives

$$
F_{n \neq 0} = -\frac{k_B T}{8\zeta^2} \sum_{n=1}^{\infty} (\Delta_{13})^2 [1 - 4\zeta^2 \ln \zeta + \cdots]. \quad (5.32)
$$

On the other hand, for the term with $n = 0$, $\xi_n = 0$, the dielectric constants of water and hydrocarbon are $\varepsilon_3(0) \approx 80$, $\varepsilon_1(0) \approx 2$, whence $\Delta_{13}(i\xi_n = 0) \approx -\frac{78}{82}$. Thus

$$
F_{n=0} \approx -\frac{k_B T}{16\zeta^2} \left(\sum_{\nu=1}^{\infty} \frac{\Delta_{13}^2(0)}{\nu^3} - 4\zeta^2 \ln \zeta [1 - \Delta_{13}(0)] \ln [1 - \Delta_{13}^2(0)] + \cdots \right)
$$
$$
\approx -\frac{k_B T}{16\zeta^2} [1 - 12\zeta^2 \ln \zeta + \cdots]. \quad (5.33)
$$

We compare the separate contributions of (5.32) and (5.33). It is clear from inspection that the perturbation expansion breaks down and goes over to the weaker $1/R^6$ form much more rapidly for the microwave terms than for the remainder. In fact, since $\zeta \propto l^{1/2}$, the contribution to the force from ultraviolet and infrared correlations retains the strong $1/l$ form over a separation distance of about 10 times that of the microwave contribution. In other words as two small oil drops come close together, the force between them is determined by the relatively weak $1/R^7$ form at first; as they come closer together the microwave contribution retains this $1/R^7$ behaviour, while the ultraviolet and infrared begin to "see" the spheres with a stronger force law $\propto 1/l^2$. At extremely close distances, the relative contribution of all frequency regimes is as for planes. This is exactly what one might have expected, as specificity of interaction must depend strongly on specific features of the ultraviolet and infrared spectra of the interacting bodies. Detailed calculations of the force magnitudes between spheres as a function of separation for various model systems such as polystyrene across water, water across polystyrene, polystyrene across heptane have been made by Smith, Mitchell and Ninham (1973).

The retarded interaction problem is much more difficult except in the region of large separations, or equivalently, for very small spheres. The mathematical difficulties have to do with the non-separability or even approximate separability of the Helmholtz equation. When the dielectric constants of the two spheres are close to those of the intervening medium, some progress can be made, and this problem has been treated at length by Langbein (1970). The conclusions of these several rather involved mathematical detours can be summed up for practical purposes rather simply however, and are as follows:

(1) Ultraviolet and infrared contributions can be handled by Hamaker theory with corrections for retardation. Thus

$$E(\text{uv}+\text{ir}) = -\frac{A(\text{eff})}{6}\left\{\frac{2b^2}{R^2-4b^2}+\frac{2b^2}{R^2}+\ln\left(1-\frac{4b^2}{R^2}\right)\right\},$$

where

$$A(\text{eff}) = k_B T \sum_{n=1}^{'n_s}\left\{\sum_{\nu=1}^{\infty}\left(\frac{\Delta_{13}^2}{\nu^3}\right)\right\} \approx \hbar \int_{\xi_1}^{\xi_s}\Delta_{13}^2\,d\xi. \tag{5.34}$$

The cut-off frequency ξ_s due to retardation is the same as that for planes

$$\xi_s = \left(\frac{2\sqrt{\varepsilon_3}l}{c}\right).$$

(2) Microwave contributions must be handled separately by formulae like those above. In the presence of salt, the microwave term will be screened with a cut-off distance of the order of the Debye length.

A more elegant expression for the non-retarded interaction energy has been given by Love (1975) who solved the problem exactly in bispherical coordinates, and the formula does not suffer the same computational difficulties as that of Langbein. He shows also that the simple approximate formulae (5.21), (5.30) are correct to $\leq 10\%$ if [(5.21)] $l/b > 1/3$ and [(5.30)]

$$|\Delta| \simeq 1 \frac{l}{b} < \frac{1}{50} \, ;$$

$$|\Delta| \simeq \frac{1}{2} \text{ no restriction;}$$

$$|\Delta| \simeq 0 \frac{l}{b} < \frac{1}{10} \, .$$

The retarded dispersion forces between small conducting spheres have been investigated by Feinberg (1974).

5.5 Cylinders

The problem of interacting rod-like particles has been considered by a number of authors, who used various different techniques (Langbein, 1972b, 1973a; Parsegian, 1972; Mitchell, Ninham and Richmond, 1972, 1973a,b; Israelachvili, 1973a). For two parallel cylinders "1" and "2" of radii b_1, b_2 and susceptibilities $\varepsilon_1, \varepsilon_2$ immersed in a medium of susceptibility ε_3 and of length L, the centre-to-centre distance R, $(L \gg R)$ the exact expression for the interaction free energy can be shown by the surface mode method (Mitchell, Ninham and Richmond, 1972, 1973) to be

$$F(R) = \frac{k_B T L}{\pi} \sum_{n=0}^{\infty}{}' \int_0^{\infty} dk \ln |\mathcal{I} - \mathbf{\Omega}|, \tag{5.35}$$

where $\mathbf{\Omega} = \mathcal{M} \mathcal{N}$ and the infinite matrices $\mathcal{M}$, $\mathcal{N}$ are defined by

$$M_{nm} = \frac{(\varepsilon_3 - \varepsilon_1) I_n'(kb_1) I_n(kb_1) K_n(kb_1) K_{n+m}(kR)}{[\varepsilon_1 I_n'(kb_1) K_n(kb_1) - \varepsilon_1 I_n(kb_1) K_n'(kb_1)] K_m(kb_2)} \tag{5.36}$$

and

$$N_{nm} = M_{nm}(\varepsilon_1 \to \varepsilon_2, \, b_1 \to b_2). \tag{5.37}$$

While (5.35) formally represents the solution to the problem, the expression is singularly uninstructive. Langbein (1972) has derived eqns. (5.35)–(5.37) by his formalism and has succeeded in recasting (5.35) into the form of a lengthy summation, as in his analysis of the sphere–sphere interaction, which must be evaluated numerically. Except for very close or very wide separations, as for spheres, the sum converges slowly. In the limit of very close separation $l = R - (b_1 + b_2) \to 0$, Langbein has also shown by a tour de force of heavy manipulation that the result (5.35) goes over to the Lifshitz expression for planes multiplied by the same geometric factor which occurs in Hamaker theory, viz.

$$E = -\frac{\hbar L}{16\pi}\left(\frac{lb_1 b_2}{2R}\right)^{1/2}\frac{1}{l^2}\sum_{\nu=1}^{\infty}\frac{1}{\nu^3}\int_0^{\infty}(\Delta_{13}\Delta_{23})^{\nu}\,d\xi. \quad (5.38)$$

At large separations b_1/R, $b_2/R \gg 1$, only a few terms of the determinant are significant, and asymptotic expressions are obtainable. We quote the result only for similar cylinders and have for the free energy per unit length

$$F(R) \approx -\left(\frac{3b^2}{4}\right)^2\frac{\pi k_B T}{2R^5}\sum_{n=0}^{\infty}{}'\left\{[\Delta^{\perp}(\Delta^{\perp}+\Delta^{\parallel})+\tfrac{3}{16}(\Delta^{\parallel}-\Delta^{\perp})^2]+\frac{\delta(R)}{R^2}\right\}, \quad (5.39)$$

where

$$\Delta^{\perp} = \frac{(\varepsilon_1^{\perp}-\varepsilon_3)}{(\varepsilon_1^{\perp}+\varepsilon_3)}, \qquad \Delta^{\parallel} = \frac{\varepsilon_1^{\parallel}-\varepsilon_3}{2\varepsilon_3} \quad (5.40)$$

depend on transverse and parallel rod polarizabilities. The correction term in $\delta(R)/R^2$ which determines how rapidly the "thin" rod approximation (5.39) breaks down as R decreases is a rather complicated expression which has been given by Mitchell, Ninham and Richmond (1973a) and will not be reproduced here. We simply note that for like cylinders $b_1 = b_2 = b$ it is proportional to $b^2[\gamma_1 + \gamma_2 \ln(b/R)]$, where γ_1 and γ_2 are independent of R. As $\Delta^{\perp}, \Delta^{\parallel} \to 0$, terms in $\ln(b/R)$ disappear, and (5.39) reduces to the expansion of the Hamaker result. Thus if ε_3 is a vacuum, and the cylinders have a single absorption frequency, $\varepsilon_1 = 1 + 4\pi N\alpha_0/1 + (\xi/\omega_0)^2$, (5.39) collapses to the form

$$E(R) \approx -\frac{9\pi^3 b^4}{32R^5}\left(1+\frac{25}{4}\frac{b_2}{R^2}+\cdots\right)\hbar\omega_0\alpha^2 N^2 \quad (5.41)$$

in agreement with (1.50, 1.54). Effects of retardation on thin rod interactions have been investigated by Mitchell, Ninham and Richmond (1973b), and damping due to retardation follows much the same pattern as that for planes, as would be expected.

If we recall the behaviour of multilayer systems discussed in Chapter 3, we should not be surprised to find that similar complications occur with arrays of cylinders—by far the most interesting system for applications in biology. A general investigation of van der Waals attraction in arrays of cylinders and spheres has been carried out by Langbein (1973a,b). Although quite general, the analysis has not yet been carried through to a form which permits easy computation. A more explicit analysis of the free energies of arrays of cylinders and of three-body forces has been carried out by Smith, Mitchell and Ninham (1974). This paper is a collector's item having the rare distinction that all three authors failed to notice the omission of a crucial page of manuscript in galley proof. The omission is catastrophic, in so far as the paper is thereby rendered incomprehensible. But some conclusions are worth noting: From (5.39) to (5.41) it can be shown that the "thin" cylinder approximation breaks down severely at a distance regime given by $b/R \sim \frac{1}{4}$. For a hexagonally close packed array it turns out that three-body forces also become comparable with the correction terms to (5.39) and (5.41) at about this distance also. This implies that for the most interesting applications (e.g., models for muscle where $b/R \approx \frac{1}{4}$), all many-body forces must be included, and assumptions of pairwise additivity of cylinder–cylinder interactions are probably invalid—particularly when water is involved. We remark incidentally that the 3-cylinder interaction tends to align cylinders in a hexagonally packed array.

For thin rods of infinite length inclined at an angle θ, the free energy has the form (Parsegian, 1972)

$$F(R, \theta) = -\frac{3b^4}{8} \frac{\pi k_B T}{(\sin \theta) R^4} \sum_{n=0}^{\infty}{}' \left\{ \Delta^\perp (\Delta^\perp + \Delta^\parallel) + \frac{1 + 2 \cos^2 \theta}{16} (\Delta^\perp - \Delta^\parallel)^2 \right\}.$$

(5.42)

Note the R^{-5} and R^{-4} dependence for parallel and skewed rods. The angular dependence of $F(R, \theta)$ implies a torque. The corresponding expression for a rod of finite length interacting with an infinite rod has been obtained by Mitchell and Ninham (1973).

5.6 Anistropy

We now extend our analysis to include effects due to anisotropy of dielectric properties. Simply because no other physical mechanism exists which will explain observed results, effects due to anisotropy must be central in accounting for many phenomena observed in liquid crystals, a point which has been stressed by de Gennes (1970). Using London–Hamaker theory de Gennes considered the interaction between a nematic or cholesteric liquid crystal droplet and an identical bulk liquid crystal phase across an isotropic medium. He showed that the droplet is attracted if its axis of anisotropy is parallel with that of the bulk phase and repelled if the two axes are perpendicular. This example illustrates the importance of anisotropy in determining the nature (and in particular the sign) of van der Waals forces. Effects of temperature must be considered since experimentally interactions are still comparatively strong over distances of 1 or 2 μm, a very much greater distance from the cohesive length of the liquid

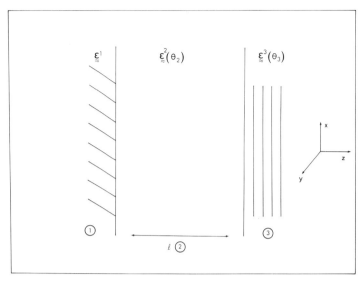

FIG. 5.8. Two anisotropic bodies "1" and "3" interacting across a planar slab of anisotropic medium "2".

crystal, and far enough so that retardation damps out ultraviolet correlations. We begin with planar systems.

Dielectric anisotropy in planar systems

Katz (1971) investigated the interaction between anisotropic planar media using the methods of quantum field theory. He confirmed the existence of a torque tending to rotate two anisotropic crystals (separated by an air gap) relative to each other. However, attention was confined to the zero temperature case. A much simpler derivation of the interaction energy has been given by Parsegian and Weiss (1972) using the surface mode method. The authors consider the case of a planar slab of material of thickness l separating two semi-infinite media (Fig. 5.8). All three media have anisotropic dielectric properties, but isotropic magnetic properties with magnetic permeability unity. Assume further that the dielectric tensor in medium r can be represented by means of principal dielectric axes as

$$\varepsilon = \begin{pmatrix} \varepsilon_x^{(r)} & 0 & 0 \\ 0 & \varepsilon_y^{(r)} & 0 \\ 0 & 0 & \varepsilon_z^{(r)} \end{pmatrix}, \qquad r = 1, 2, 3. \tag{5.43}$$

Finally, make the assumption that the third dielectric axis of each medium coincides with the z-axis. This allows the dielectric axes to be characterized by two angles θ_2, θ_3 representing the rotation of materials 2 and 3 with respect to material 1 about the z-axis. The case $\theta_2 = \theta_3 = 0$ refers to the situation in which the principal dielectric axes of all the media are parallel. If the assumption that the "3" axes are identical does not obtain, the analysis can be carried through as below, but the algebra proliferates.

The effects of rotating materials 2 and 3 by θ_2, θ_3 can be represented by the use of dielectric tensors $\varepsilon^2(\theta_2)$, $\varepsilon^3(\theta_3)$, where

$$\boldsymbol{\varepsilon}^r(\theta_r) = \begin{pmatrix} \varepsilon_x^{(r)} + (\varepsilon_y^{(r)} - \varepsilon_x^{(r)}) \sin^2 \theta_r & (\varepsilon_x^{(r)} - \varepsilon_y^{(r)}) \sin \theta_r \cos \theta_r & 0 \\ (\varepsilon_x^{(r)} - \varepsilon_y^{(r)}) \sin \theta_r \cos \theta_r & \varepsilon_y^{(r)} + (\varepsilon_x^{(r)} - \varepsilon_y^{(r)}) \sin^2 \theta_r & 0 \\ 0 & 0 & \varepsilon_z^{(r)} \end{pmatrix}. \quad (5.44)$$

Now in the non-retarded limit, Maxwell's equations do not reduce to the Laplace equation, but rather to

$$\nabla \cdot (\boldsymbol{\varepsilon} \nabla \phi) = 0, \qquad \mathbf{E} = -\nabla \phi. \quad (5.45)$$

This equation is to be solved with the boundary conditions that E_x, E_y, $D_z = (\boldsymbol{\varepsilon}\mathbf{E})_z$ be continuous at the interfaces. Putting

$$\phi_r = f_r(z) \exp[i(ux + uy)]; \qquad r = 1, 2, 3 \quad (5.46)$$

and substituting into (5.45) leads to the equation

$$\varepsilon_z^{(r)} f_r''(z) - (\varepsilon_{11}^{(r)} u^2 + 2\varepsilon_{12}^{(r)} uv + \varepsilon_{22}^{(r)} v^2) f_r(z) = 0, \quad (5.47)$$

where $\varepsilon_y^{(r)}$ is the y element of the rotated dielectric tensor (5.44). The solutions to (5.47) are

$$f_r(z) = A_r e^{\beta_r z} + B_r e^{-\beta_r z} \quad (5.48)$$

with

$$\beta_r(z) = \frac{\varepsilon_x^{(r)}}{\varepsilon_z^{(r)}} (u \cos \theta + v \sin \theta)^2 + \frac{\varepsilon_y^{(r)}}{\varepsilon_z^{(r)}} (v \cos \theta - u \sin \theta)^2, \quad (5.49)$$

and since only surface modes are required $A_3 = B_1 = 0$. In terms of $f_r(z)$ the boundary conditions at $z = 0$ and $z = l$ are

$$\begin{aligned} f_1(0) &= f_2(0), & \varepsilon_z^{(1)} f_1'(0) &= \varepsilon_z^{(2)} f_2'(0) \\ f_2(l) &= f_3(l), & \varepsilon_z^{(2)} f_2'(l) &= \varepsilon_z^{(3)} f_3'(l) \end{aligned} \quad (5.50)$$

whence the dispersion relation follows as

$$D(l) = 1 - \Delta^2(\omega, u, v, \theta) e^{-2\beta_2(\theta_2)l} = 0, \quad (5.51)$$

where

$$\Delta^2 = \left(\frac{a-1}{a+1}\right)\left(\frac{b-1}{b+1}\right), \qquad a = \frac{\varepsilon_z^{(1)} \beta_1}{\varepsilon_z^{(2)} \beta_2(\theta_2)}, \qquad b = \frac{\varepsilon_z^{(3)} \beta_3(\theta_3)}{\varepsilon_z^{(2)} \beta_2(\theta_2)}. \quad (5.52)$$

The free energy of interaction follows as

$$F(l) = \frac{k_B T}{4\pi^2} \sum_{n=0}^{\infty}{}' \int\int_{-\infty}^{\infty} du\, dv\, \ln D(i\xi_n; u, v). \tag{5.53}$$

Introducing polar coordinates $u = \rho \cos \phi$, $v = \rho \sin \phi$, $\beta_r(\theta)$ can be written

$$\beta_r^2(\theta) = \frac{\rho^2}{\varepsilon_2^{(r)}} (\varepsilon_x^{(r)} \cos^2 (\theta - \phi) + \varepsilon_y^{(r)} \sin^2 (\theta - \phi)) \equiv \rho^2 g_r^2(\theta - \phi) \tag{5.54}$$

and $F(l)$ takes the form

$$F(l) = \frac{k_B T}{4\pi^2} \sum_{n=0}^{\infty}{}' \int_0^{2\pi} d\rho \int_0^{\infty} \rho\, d\rho\, \ln (1 - \Delta_n^2(\theta, \phi) e^{-2\rho g_2(\theta-\phi)l}). \tag{5.55}$$

Parsegian and Weiss have carried out several model calculations to quantify the effects of anisotropy. For example, if media 1 and 3 are uniaxial crystals of HgCl (which has a very high degree of dielectric anisotropy) interacting across a layer of melt whose dielectric properties are the average of the anisotropic principal susceptibilities, the free energy is

$$F(l) \approx -\frac{\hbar}{32\pi^3 l^3} S(\theta), \tag{5.56}$$

where the value of $S(\theta)$ ranges from approximately $6 \times 10^{14}\, \text{sec}^{-1}$ at $\theta = 0$ to $-6 \times 10^{14}\, \text{sec}^{-1}$ at $\theta = 90°$. Thus the force changes from attractive to repulsive as θ increases from 0 to 90°. Parsegian and Weiss also find that the relative effects of anisotropy are very much smaller if the intermediate medium does not have the special property of "looking" dielectrically like an average of the dielectric properties of the anisotropic medium.

For a liquid crystalline medium, the isotropic phase often does have a dielectric constant which is a mean of the anisotropic constants in the nematic phase. Thus one expects comparatively strong effects due to anisotropy in the interaction of an anisotropic crystal or nematic liquid crystal across a layer of liquid crystalline material in its isotropic phase. A further point of interest in the application of the theory of van der Waals forces to liquid crystalline materials arises because nematic liquid crystals have very small elastic moduli. This means that the action of van der Waals forces on a liquid crystalline medium may well involve a distortion of the ordered structure of the liquid crystal. Theoretical predictions of the amount of such distortion for a special model system have been made by Smith and Ninham (1973) in connection with an experimental program launched by Guyon. Experiments to measure the elastic response of liquid crystals to van der Waals forces would be significant for the behaviour of many membraneous systems—all biological membranes are basically liquid crystalline in nature and most cell–cell interactions are at least in part mediated by van der Waals forces, so we may expect that the response of cell membranes to interactions with other cells or to foreign bodies may involve a distortion of the liquid crystalline order. A further application of the theory involving an applied magnetic torque has been made by Richmond and White (1974).

Anisotropy in interaction of cylinders

The longitudinal component of the polarizability tensor of a CH_2 group in a hydrocarbon chain is significantly larger than the transverse component. The effect of this anisotropy has been studied by Zwanzig (1963) who calculated by perturbation theory the interaction between parallel linear chains of coupled Drude oscillators. An extension of the surface mode analysis to deal with anisotropic parallel cylinders has been made by Mitchell, Ninham and Richmond (1972). In the cross-sectional plane the cylinders are assumed to be isotropic, with a different susceptibility in the z direction so that

$$\mathbf{\varepsilon}_i = \begin{pmatrix} \varepsilon_i^t & 0 & 0 \\ 0 & \varepsilon_i^t & 0 \\ 0 & 0 & \varepsilon_i^l \end{pmatrix} \tag{5.57}$$

and the surface modes are determined by the modified Laplace equation

$$\nabla \cdot (\mathbf{\varepsilon} \nabla \phi) = 0 \tag{5.58}$$

with appropriate boundary conditions. The analysis is much the same as that required for §5.5, and if "1", "2" denote cylinders of radii b_1, b_2 and "0" denotes the intervening medium the result in the thin cylinder approximation $b_1, b_2 \ll R$ where R is the centre-to-centre distance is

$$F(b_1, b_2, R) = -\frac{\pi L b_1^2 b_2^2}{2^{11} R^5} k_B T \sum_{n=0}^{\infty} {}' \left\{ 27 \frac{(\varepsilon_0 - \varepsilon_1^l)(\varepsilon_0 - \varepsilon_2^l)}{\varepsilon_0^2} \right.$$

$$+ 684 \frac{(\varepsilon_0 - \varepsilon_1^t)(\varepsilon_0 - \varepsilon_2^t)}{(\varepsilon_0 + \varepsilon_1^t)(\varepsilon_0 + \varepsilon_2^t)}$$

$$\left. + 90 \left[\frac{(\varepsilon_0 - \varepsilon_1^l)}{\varepsilon_0} \left(\frac{\varepsilon_0 - \varepsilon_2^t}{\varepsilon_0 + \varepsilon_2^t} \right) + (\varepsilon_1 \leftrightarrow \varepsilon_2) \right] \right\}. \tag{5.59}$$

In the limit that $\varepsilon^t = \varepsilon^l$, i.e., the cylinders are isotropic, this result coincides with (5.39). In the special case that the intervening medium is a vacuum, $T = 0$, and each polarizable group of the cylinder exhibits Lorentz oscillator dispersion at a single frequency (5.59) reduces to the result of Zwanzig. The essential point of his calculation was to demonstrate the importance of non-additivity and anisotropy in the interaction of long chain molecules, and by comparing his result with the earlier work of Salem (1962) who ignored these effects, Zwanzig showed that anisotropy and non-additivity due to coupling between Drude oscillators both gave significant contributions to the energy of interaction of hydrocarbon chains. The main new features which emerge from the more general result (5.59) when the intervening medium is water are that, (1) the zero frequency entropic contributions to the free energy are of the same order of magnitude as ultraviolet contributions. This is not surprising. (2) When the intervening medium is a vacuum Zwanzig found that the dominant contributions to the energy arise from contributions from the longitudinal susceptibility ε^l. For hydrocarbon chains in water, the opposite conclusion holds—terms in ε^t being more important. This is interesting since it implies that local changes in geometry and side chains give significant contributions to the free energy which is intuitively what we would expect. The corresponding free energy for crossed anisotropic cylinders has been derived by Mitchell and Ninham (1973) by the formalism of Langbein.

Forces between anisotropic and asymmetric molecules

The role of an anisotropy in polarizability in aligning small molecules near a surface or in close proximity with each other, and the interaction between ellipsoidal particles has been investigated by Imura and Okano (1973) and by Israelachvili (1973a). For further discussion of effects due to anisotropy and asymmetry and their possible relevance in biology, e.g., as a steering mechanism in aligning the active sites of proteins and other macromolecules as they come together, see Israelachvili (1974), Israelachvili and Tabor (1973).

5.7 The method of Parsegian

A method for deducing the forces between long anisotropic thin rods has been devised by Parsegian (1972). This method is ingenious. Because its potential for investigating van der Waals forces in arrays or condensed media has not been exploited in full, it deserves separate exposition. Essentially, this technique is based on the very sound principle that if the answer to a problem can not be easily found, a careful perusal of the works of Lord Rayleigh often provides the missing key.

Following Parsegian (1972), we imagine a composite system of a continuous phase of substance in which are embedded infinitely long parallel cylindrical rods of material r and radius b; the rods are packed at an average cross-sectional density of N rods per unit area. Let the dielectric susceptibility ε_m of the embedding medium m be isotropic, while the susceptibility of the rod substance is $\varepsilon_r^{\parallel}$ and $\varepsilon_r^{\perp}$ parallel and perpendicular to the direction of the uniaxial symmetry. The net susceptibility along the principal axis, parallel to the rods is

$$\varepsilon^{\parallel} \equiv \alpha = \varepsilon_m + v(\varepsilon_r^{\parallel} - \varepsilon_m), \tag{5.60}$$

where $v = N\pi b^2 < 1$ is the volume fraction of rod material.

Perpendicular to the rods, when their separation is much greater than their diameter, the composite susceptibility is (Lord Rayleigh, 1892)

$$\varepsilon^{\parallel} = \beta = \varepsilon_m + 2v\varepsilon_m \frac{\Delta_{rm}^{\perp}}{(1 - v\,\Delta_{rm}^{\perp})}, \tag{5.61}$$

where

$$\Delta_{rm}^{\perp} = \frac{(\varepsilon_r^{\perp} - \varepsilon_m)}{(\varepsilon_r^{\perp} + \varepsilon_m)}. \tag{5.62}$$

Equation (5.60) is an obvious result, but eqn. (5.61) is not, and is the key to the method. Knowledge of the susceptibility of a composite medium in terms of its components implicitly tells us everything we need to know about interactions. The problem of determination of the susceptibility, or equivalently thermal or electrical conductivity of a heterogeneous substance is mathematically equivalent to solving Laplace's equation for the composite. The solution has much wider application than here indicated, e.g., to geophysics, chemical engineering and vision research, biological microscopy, and has exercised some formidable intellects including Maxwell. For reviews which also contain some useful references, see Meredith and Tobias (1962), Dukhin and Shilov (1974). Lord Rayleigh (1892) was the first to make substantial progress, and not a great deal more has been done since, see, e.g., Ninham and Sammut (1976). For a square array of parallel cylinders, one has

$$\varepsilon^{\perp} = \varepsilon_m + 2v\varepsilon_m\,\Delta_{rm}^{\perp}/(1 - v\,\Delta_{rm}^{\perp} + 0{\cdot}306\,\Delta_{rm}^2 v^4 + 0{\cdot}133\,\Delta_{rm}^2 v^8 + \cdots). \tag{5.63}$$

The influence of rod arrangement shows up only in terms of order v^5 and higher. The last term in v^8 was computed by Runge (1925) by Rayleigh's method. Runge also gives the corresponding result for concentric shells.

Rayleigh's main motivation was the justification of the famous Lorentz-Lorenz formula expressing the connection between refractive index and density for spherical particles. For spheres of susceptibility ε_s and radius b embedded in a medium ε_m the composite susceptibility ε_c is given by

$$\frac{\varepsilon_c - \varepsilon_m}{\varepsilon_c + 2\varepsilon_m} = p\,\frac{\varepsilon_s - \varepsilon_m}{\varepsilon_s + 2\varepsilon_m}; \qquad p = \frac{4\pi}{3}\frac{b^3}{R^3}, \tag{5.64}$$

where R is the mean centre-to-centre distance of the spheres. In lowest approximation this is equivalent to the Clausius-Mossotti relation, or for $\varepsilon_c \sim 1$, $\varepsilon_m = 1$, to the result $\varepsilon = 1 + 4\pi N\alpha$ which we have used so often. Rayleigh's work provides an estimate for the range of validity of this expression. He gives (a numerical error has been corrected)

$$\varepsilon_c = 1 + \frac{3p}{\left(\dfrac{\varepsilon_s + 2\varepsilon_m}{\varepsilon_s - 1} - p - \dfrac{0 \cdot 592(\varepsilon_s - \varepsilon_m)p^{10/3}}{(\varepsilon_s + \frac{4}{3}\varepsilon_m)} + \cdots\right)}. \tag{5.65}$$

Again the influence of random ordering or arrangements different from cubical shows up in higher order in p. This is an important result. The denominator is an asymptotic expression which will be useful for computation when

$$\frac{0 \cdot 592(\varepsilon_s - \varepsilon_m)p^{10/3}}{\varepsilon_s + \frac{4}{3}\varepsilon_m} \le p. \tag{5.66}$$

In the worst case $\varepsilon_m = 1$, taking $\varepsilon_s \approx 2$ as typical of most organic substances and water in the ultraviolet, we require for use of the formula $p \lesssim 2 \cdot 1$. For close packing of spheres (or in this case atoms) p has the maximum value $\pi/6 \sim \frac{1}{2}$, and the inequality is well satisfied. Rayleigh's formula then shows why the Clausius–Mossotti approximation is generally such a good approximation for non-polar substances. Although Rayleigh's method of derivation was quite general, he confined his explicit analysis to spheres in cubical arrays, or cylinders in square arrays. Surprisingly his method has not been extended to other arrays. We give one such example. For an array of cylinders infinite in the z direction, with spacing c in the x direction and d in the y direction, the composite susceptibility can be shown to be (Ninham and Sammut, 1976)

$$\varepsilon_{cx} = \varepsilon_m \left\{ 1 + \frac{2\pi b^2}{cd} \frac{\Delta_{rm}^\perp}{\left(1 - \dfrac{b^2}{c^2}\Delta_{rm}^\perp S_2 + \dfrac{3b^8}{c^8} S_4^2 (\Delta_{rm}^\perp)^2 + \cdots\right)} \right\}, \tag{5.67}$$

where

$$S_2 = \tfrac{4}{3}\{3KE - (2 - k^2)K^2\} \tag{5.68}$$

$$S_4 = \frac{16}{45} K^4[K^4 - K^2 + 1], \tag{5.69}$$

where $K = K(k)$, $E = E(k)$ are complete elliptic functions of the first and second kind, with moduli k determined through $K'/K = d/c$. K' is the elliptic function with complementary modulus $k' = \sqrt{1 - k^2}$. The coefficients S_2 and S_4 are easily evaluated with the aid of the theory of theta

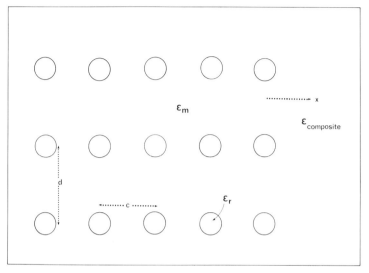

FIG. 5.9. Illustrating eqns. (5.84) and (5.85).

functions. Thus, when $c \ll d$ (Fig. 5.9), (5.67) reduces to

$$\varepsilon_{cx} \approx \varepsilon_m \left\{ 1 + \frac{2v\,\Delta^{\perp}_{rm}}{\left(1 - \frac{v\pi}{3}\frac{d}{c}\Delta^{\perp}_{rm} - \frac{(\Delta^{\perp}_{rm})^2 v^4}{675}\left(\frac{\pi d}{c}\right)^4 + \cdots\right)} \right\} \qquad (5.70)$$

while if $d \ll c$,

$$\varepsilon_{cx} \approx \varepsilon_m \left\{ 1 + \frac{2v\,\Delta^{\perp}_{rm}}{\left(1 + v\Delta^{\perp}_{rm}\frac{\pi}{3}\frac{c}{d}\left(1 - \frac{6}{\pi}\frac{d}{c}\right) - \frac{v^4 \Delta^2_{rm}}{675}\left(\frac{\pi c}{d}\right)^4 + \cdots\right)} \right\}$$

$$(5.71)$$

where

$$v = \frac{\pi b^2}{cd}.$$

When $c \ll d$, or $d \ll c$, the arrays begin to "look like" stacks of plates. In the first case, comparison of (5.70) with standard results in optics (Born and Wolf, 1965) shows that the effective volume fraction if the cylinders are regarded as effective plates of susceptibility ε_r is

$$f_1 = \frac{2v\varepsilon_m}{(\varepsilon_r + \varepsilon_m)\left(1 - \frac{\pi}{3}v\Delta_{rm}\frac{d}{c} - \frac{v^4 \Delta^4_{rm}}{675}\left(\frac{\pi d}{c}\right)^4 + \cdots\right)} \qquad (5.72)$$

while in the second perpendicular to the "plates," the effective volume fraction of "plates" is

$$f_1 = \frac{2v\varepsilon_r}{(\varepsilon_r + \varepsilon_m)\left(1 + 2v\Delta_{rm} + \frac{v\pi}{3}\frac{c}{d}\Delta_{rm}\left(1 - \frac{6}{\pi}\frac{d}{c}\right) + \cdots\right)}. \qquad (5.73)$$

Formulae like those above should be useful in examining the phase transitions which occur in soap solutions, liquid crystalline phenomena and a variety of biological situations. Their potential for such studies has not yet been exploited.

The result of Parsegian and Weiss (1972) for the van der Waals interaction of two semi-infinite anisotropic media can now be used to write down the interaction of two composite media. Let this interaction occur across a planar state of material m and thickness l, (Fig. 5.8), where l is much greater than the spacing between rods in each medium. The two sets of embedded rods are parallel to the slab face, but make an angle θ with each other (Fig. 5.10). On the left the rods are parallel to the x direction while on the right they make an angle θ with the x direction. Neglecting retardation, it follows from eqn. (5.51) that the free energy of

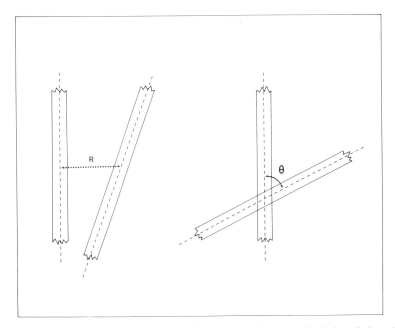

FIG. 5.10. Rods separated by a minimum distance R, making an angle θ viewed along R.

interaction with respect to infinite l is

$$F(l, \theta) = -\frac{k_B T}{16\pi^2 l^2} \sum_{n=0}^{\infty}{}' \int_0^{2\pi} d\phi \sum_{n=1}^{\infty} \frac{\Delta_{mr}(\phi) \Delta_{mr}(\theta - \phi)}{n^3}, \quad (5.74)$$

where

$$\Delta_{mr}(\theta) = \frac{\varepsilon_m - \beta(1 + \gamma \cos^2 \theta)^{1/2}}{\varepsilon_m + \beta(1 + \gamma \cos^2 \theta)^{1/2}} \quad (5.75)$$

with $\gamma = (\alpha - \beta)/\beta$. To proceed further, we expand Δ_{mr} in powers of v and retain only leading terms. This gives

$$\Delta_{mr}(\theta) \Delta_{mr}(\theta - \phi) = v^2 \left\{ \left[(\Delta_{rm}^\perp)^2 + \tfrac{1}{4} \Delta_{rm}^\perp \left(\frac{\varepsilon_r^\| - \varepsilon_m}{\varepsilon_m} - 2\Delta_{rm}^\perp \right) \right] \right.$$
$$\times [\cos^2 \theta + \cos^2(\theta - \phi)] + \frac{1}{16} \left(\frac{(\varepsilon_r^\| - \varepsilon_m)}{\varepsilon_m} - 2\Delta_{rm}^\perp \right)^2$$
$$\left. \times \cos^2 \theta \cos^2(\theta - \phi) \right\} + 0(v^3). \quad (5.76)$$

Expansion in this form is permissible provided

$$|\gamma| \approx v |(\varepsilon_r^\| - \varepsilon_m)/\varepsilon_m - 2\Delta_{rm}^\perp| \ll 1. \quad (5.77)$$

To the same approximation in v, the free energy of interaction is then given by

$$F(l, \theta) = -\frac{k_B T}{8\pi l^2} N^2 (\pi b^2)^2 \sum_{n=0}^{\infty}{}' \left\{ (\Delta_{rm}^\perp)^2 + \tfrac{1}{4}(\Delta_{rm}^\perp) \left(\frac{\varepsilon_r^\| - \varepsilon_m}{\varepsilon_m} - 2\Delta_{rm}^\perp \right) \right.$$
$$\left. + \frac{2\cos^2 \theta + 1}{2^7} \left(\frac{\varepsilon_r^\| - \varepsilon_m}{\varepsilon_m} - 2\Delta_{rm}^\perp \right)^2 \right\}. \quad (5.78)$$

The factor N^2 implies that to leading order in $v = N(\pi b^2)$ the total interaction between the two composite media is the pairwise sum of individual rod–rod interactions. This expression for $F(l, \theta)$ implicitly contains all many-body properties of the interaction between individual rods via the experimental $\varepsilon_r^\|$, $\varepsilon_r^\perp$ and ε_m.

The individual rod-rod interaction follows from an argument identical with that used by Pitaevskii (1959) in his derivation of the pairwise interaction of solute molecules in dilute solution. Thus let $f^\|(\rho)$ be the free energy per unit length between two parallel rods, and $f(\rho, \theta)$ per pair of infinitely long skewed rods; ρ is their nearest interaxial distance and θ their mutual angle. If the parallel rods point in the x direction, then we must have

$$F(l, \theta) = \frac{v^2}{(\pi b^2)^2} \int_0^\infty dz' \int_0^\infty dz \int_{-\infty}^\infty dy \, f^\parallel(\rho), \tag{5.79}$$

where $\rho = (l + z + z')^2 + y^2$, while for skewed rods

$$F(l, \theta) = \sin \theta \left(\frac{v}{\pi b^2}\right)^2 \int_0^\infty dz' \int_0^\infty dz f(\rho, \theta). \tag{5.80}$$

Then twice differentiating with respect to l, we have

$$F^\parallel(l, 0) = \frac{v^2}{(\pi b^2)^2} \int_{-\infty}^\infty f^\parallel(\rho) \, dy, \qquad \rho^2 = l^2 + y^2 \tag{5.81}$$

for parallel rods or

$$F^\parallel(l, \theta) = \sin \theta \left(\frac{v}{\pi b^2}\right)^2 f(l, \theta) \tag{5.82}$$

for skewed rods. Equation (5.82) gives the result for cylinders at an angle immediately, while by solution of the integral equation (5.81) or by direct verification by substitution (5.59) also follows.

The chief virtue of Parsegian's method is that it provides a criterion for deciding when the van der Waals interaction will be additive, i.e., when the whole free energy can be considered as a sum of rod–rod interactions. When the dielectric susceptibility of a region is itself an additive function of the materials in that region, then the van der Waals interaction will also be additive. Thus an array of rods will interact additively where the susceptibility of the whole array is proportional to the rod concentration. This is certainly so for the susceptibility parallel to the rods (5.60). But perpendicular to the rods the susceptibility is given by (5.61), and a necessary condition if the rods are to interact pairwise is that terms in v^2 or higher be negligible compared with the term in v, i.e., that $|v \, \Delta_{rm}| \ll 1$. For $\varepsilon_r = 2$, $\varepsilon_m = 1 \cdot 8$ (hydrocarbon and water in the visible region) this condition is $v \ll |\Delta_{rm}|^{-1} \approx 20$ which is easily satisfied since the maximum possible value of v is $\pi/4$ for a square array. But if $\varepsilon_r = 2$, $\varepsilon_m = 80$ (hydrocarbon and water at the limit of zero frequency), the stronger condition $v \ll 1$ must hold for additivity of the forces. For lipid cylinders in water observed as the "middle" phase of fatty acids in water, as pointed out by Parsegian (1972) the volume fraction of lipid is in the range $0 \cdot 35 \lesssim v \lesssim 0 \cdot 6$, and the inequality $v \, \Delta_{rm} \ll 1$ is strongly violated at the low frequency end of the absorption spectrum. A similar situation exists with arrays of tobacco mosaic virus where high volume fractions are observed and the protein core is likely to have a dielectric constant much lower than 80.

A second restriction to which we shall return follows from perusal of (5.77). If ionic and or electronic current fluctuations along the rods are allowed $\varepsilon_r^{\parallel}$ can become large. The parameter γ does not necessarily satisfy $\gamma \ll 1$, and the interaction between composite media cannot be decomposed into rod–rod interactions. The resulting interaction can be of much longer range. These conclusions are not definite, and will be altered strongly in the presence of salt, but they do serve a note of caution for attempts to deduce the properties of regular assemblies from two-body interactions.

The method of Parsegian had also been used by Richmond (1974) to study the van der Waals free energy for a rod of finite length immersed in a uniform electrolyte solution and inclined at an arbitrary angle with respect to a plane dielectric substrate. Expressions for the torque exerted on the rod and the spring constant for torsional oscillations of the rod due to surface interactions are also given. For adsorbed rod-like molecules these oscillation frequencies should be observable using spectroscopic methods.

References

Born, M. and Wolf, E. (1965). "Principles of Optics," 3rd ed. Pergamon, London. p. 705ff.

de Gennes, P. G. (1970). *CR. Acad. Sci.* **271,** 469.

Dukhin, S. S. and Shilov, V. N. (1974). "Dielectric Phenomena and the Double Layer in Disperse Systems and Polyelectrolytes." John Wiley & Sons, New York.

Dzyaloshinskii, I. E., Lifshitz, E. M. and Pitaevskii, L. D. (1961). *Advan. Phys.* **10,** 165.

Feinberg, G. (1974). *Phys. Rev. B* **9,** 2490.

Forsyth, Jr., P. A., Marčelja, S., Mitchell, D. J. and Ninham, B. W. (1977). *J. Chem. Soc., Farad. II* (to appear).

Hamaker, H. C. (1937). *Physica* **4,** 1058.

Huisman, F. and Mysels, K. J. (1969). *J. Phys. Chem.* **73,** 489.

Imura, H. and Okano, K. (1973). *J. Chem. Phys.* **58,** 2763.

Israelachvili, J. N. (1973a). *J. theoret. Biol.* **42,** 411.

Israelachvili, J. N. (1974). *Quart. Rev. Biophys.* **6,** 341

Israelachvili, J. N. and Tabor, D. (1972). *Proc. Roy. Soc. A* **331,** 19.

Israelachvili, J. N. and Tabor, D. (1973). *Prog. Surface and Membrane Sci.* **7,** 1.

Katz, E. I. (1971). *Sov. Phys. JETP* **33,** 634.

Langbein, D. (1969). *J. Adhesion* **1,** 237.

Langbein, D. (1970). *Phys. Rev. B* **2,** 3371.

Langbein, D. (1971a). *J. Phys. Chem. Solids* **32,** 133.

Langbein, D. (1971b). *J. Phys. Chem. Solids* **32,** 1654.

Langbein, D. (1972a). *J. Adhesion* **3,** 213.

Langbein, D. (1972b). *Phys. Kondens. Materie* **15,** 61.

Langbein, D. (1973a). *Adv. Solid State Phys.* (*Festkörperprobleme*) **XIII,** 85 (Pergamon Vieweg).

Langbein, D. (1973b). *J. Phys. A* **6,** 1149.

Langmuir, I. (1938). *J. Chem. Phys.* **6,** 873.

Love, J. D. (1975). *Quart. J. Mech. Appl. Maths.* **28,** 449. See also *J. Chem. Soc., Farad. II.*

Lyklema, J. and Mysels, K. J. (1965). *J. Am. Chem. Soc.* **87,** 2539.

Marčelja, S., Mitchell, D. J. and Ninham, B. W. (1976). *Chem. Phys. Letts.* (in press).

Meredith, R. E. and Tobias, C. W. (1962). *Advan. Electrochem. Eng.* **2,** 15.

Mitchell, D. J. and Ninham, B. W. (1972). *J. Chem. Phys.* **56,** 1117.

Mitchell, D. J. and Ninham, B. W. (1973). *J. Chem. Phys.* **59,** 1246.

Mitchell, D. J. and Ninham, B. W. and Richmond, P. (1972). *J. theoret. Biol.* **37,** 251.

Mitchell, D. J., Ninham, B. W. and Richmond, P. (1973a). *Biophys. J.* **13,** 359.

Mitchell, D. J., Ninham, B. W. and Richmond, P. (1973b). *Biophys. J.* **13,** 370.

Ninham, B. W. and Parsegian, V. A. (1970a). *J. Chem. Phys.* **52,** 4578.

Ninham, B. W. and Parsegian, V. A. (1970b). *J. Chem. Phys.* **53,** 3398.

Ninham, B. W. and Sammut, R. A. (1976). *J. theoret. Biol.* **56,** 125.

Onsager, L. (1941). *Ann. N.Y. Acad. Sci.* **51,** 627.

Parsegian, V. A. (1972). *J. Chem. Phys.* **56,** 4393.

Parsegian, V. A. (1973). *Ann. Rev. Biophys. Bioeng.* **2,** 221.

Parsegian, V. A. and Brenner, S. L. (1976). *Nature* **259,** 632.

Parsegian, V. A. and Gingell, D. (1972). *J. theoret. Biol.* **36,** 41.

Parsegian, V. A. and Gingell, D. (1973). "Recent Advances in Adhesion," (ed. H. L. Lee). Gordon and Breach, London. pp. 153–192.

Parsegian, V. A. and Ninham, B. W. (1971). *J. Coll. Interface Sci.* **37,** 332.

Parsegian, V. A. and Ninham, B. W. (1972). *J. theoret. Biol.* **38,** 101.

Parsegian, V. A. and Weiss, G. H. (1972). *J. Adhesion* **3,** 314.

Perram, J. W. and Smith, E. R. (1973). *Nature Physical Science* **241,** 133.

Pitaevskii, L. P. (1959). *J. Exp. Theor. Phys.* (*USSR*) **37,** 577; (1960) *Sov. Phys. JETP* **10,** 408. See also Kemoklidze, M. P. and Pitaevskii, L. O. (1971). *Sov. Phys. JETP* **32,** 1183.

Rayleigh (Lord) [J. W. Strutt] (1892). *Phil. Mag.* **34,** 481. See also Collected Works.

Richmond, P. (1974). *J. Chem. Soc., Farad. II* **70,** 229.

Richmond, P. and White, L. R. (1974). *J. Mol. Crystals and Liquid Crystals* **27,** 217.

Runge, I. (1925). *Z. Tech. Phys.* **6,** 61.

Salem, L. (1962). *J. Chem. Phys.* **37,** 2100.

Smith, E. R., Mitchell, D. J. and Ninham, B. W. (1973a). *J. Coll. Interface Sci.* **45,** 55.

Smith, E. R., Mitchell, D. J. and Ninham, B. W. (1973b). *J. theoret. Biol.* **41,** 149.

Smith, E. R. and Ninham, B. W. (1973). *Physica* **66,** 111.

Zwanzig, R. (1963). *J. Chem. Phys.* **39,** 2251.

Chapter 6

Role of the Dispersion Force Field in Adsorption and Polymer Solutions

6.1 Dispersion forces in adsorption

The concept of dispersion self-energy of a molecule near an interface finds a particularly useful role in the study of adsorption phenomena (Mahanty and Ninham, 1974). The theoretical framework for analysis of adsorption is well known (Young and Crowell, 1962; de Boer, 1968) and needs no detailed exposition here. As we have seen in §3.5 the theory for macroscopic adsorbed films can be developed on the basis of a continuum model in which the chemical potential of an adsorbate molecule is computed by summing its interaction with all the adsorbent molecules (Halsey, 1948; Cook, 1948; Dole, 1948; Landau and Lifshitz, 1958). If the dominant interaction term is due to the London–van der Waals force, the adsorption isotherm follows from eqns. (3.49) and (3.52) as

$$\ln \frac{p}{p_0} = -\frac{A}{6\pi k_B T l^3},$$ (6.1)

where A is the Hamaker constant, l the thickness of the adsorbed film, p the pressure and p_0 the saturated vapour pressure of the adsorbate. The work of Dzyaloshinskii, Lifshitz and Pitaevskii (1961) provides a basis for connecting the parameter A with the bulk dielectric properties of the adsorbed film and substrate in a manner which is much more direct than is implied in the earlier work, and includes besides the effect of retarded forces. In fact the Lifshitz theory goes considerably beyond the earlier theories in so far as it is not limited to considerations of dispersion forces alone, but also of the effect of polar molecules in the system. While in the past literature the form of the adsorption isotherm for thick layers has been in dispute, the experiments of Anderson and Sabisky (1973) and comparison with theory §3.3 leave little doubt that the third power law dependence on l of eqn. (6.1), and the fourth power law dependence for the retarded region provides a correct and quantitatively precise description of many-layer adsorption for liquid helium films. Additional evidence (§3.5) comes from the predicted wetting properties of hydrocarbon films on water. While thick (on a molecular scale) films of a dielectric in equilibrium with its vapour adsorbed on a plane dielectric substrate seem then to be well understood theoretically, there is a basic difficulty in extending the continuum theory to very thin layers. This is because the dispersion force on an adsorbate molecule diverges as $1/l^3$ when it is brought near an interface. (Physically one knows (Landau and Lifshitz, 1958) that whatever be the variation of the chemical potential $\mu(l)$ for l in the range of molecular dimensions, it must tend to $-\infty$ for $l \rightarrow 0$ as $\mu \sim \ln l$, which corresponds to a weak solution of adsorbed molecules on the surface.) For very thin layers the continuum theory breaks down, and the appropriate theoretical framework must be provided by a microscopic theory such as that of Brunauer, Emmett and Teller (1938), or various modifications and improvements thereof (Young and Crowell, 1962; Brunauer, Copeland and Kantro, 1967). The important quantity which enters into such theories is the molecular partition function $\exp(-E_j/k_BT)$, where E_j is the energy needed to bring the adsorbate molecule to the gas phase from the j-th adsorbed layer. Because of the divergence of the dispersion energy of a point molecule when it is near the surface the molecular partition function due to dispersion interactions could not be defined for small j, and has been treated in these theories as a phenomenological term, thus severely limiting their predictive capabilities. The B.E.T. theory gives $\ln p/p_0 \propto 1/l$ which is impossible to reconcile with the Lifshitz isotherm eqn. (6.1). From a practical point of view, the B.E.T. theory has seen good service in multimolecular adsorption work, and it is important to understand why the theory works at all, and the limits of validity of the Lifshitz isotherm for very thin adsorbed layers.

As shown in Chapter 4, the finite size of a molecule leads to a finite value for the dispersion energy at an interface, so that the divergence difficulty disappears. We can then construct a parameter-free theory of adsorption which does reconcile the two theories, and can give a correct description of thin layers and of monomolecular adsorption.

6.2 Dispersion energies of adsorbed molecules

Our starting point is the theory of multimolecular adsorption in the form given by de Boer (1968). Figure 6.1 represents schematically the picture of adsorbed layers, with the molecule M_1 sitting on the adsorbing surface and the molecules M_2, M_3, M_4, ... sitting on monomolecular, bimolecular, trimolecular ... layers of thickness $l_j = 2(j-1)a$; $j = 1$, 2, 3, ... The principal assumptions of our approach are as follows. (i) The potential energy of M_1 is the dispersion self-energy of the molecule of radius a at the planar interface (Fig. 6.2) separating the adsorbent of dielectric constant ε_1 from the gas containing the adsorbate molecules of dielectric constant ε_3. (This energy is measured from a zero of energy which is the self-energy of a molecule in the gas phase at infinite distance from the adsorbing surface.) (ii) The potential energy of a molecule M_j ($j > 1$) in the j-th layer is the dispersion self-energy of the molecule at the second of the two planar interfaces (Fig. 6.2) separating three dielectric media of dielectric constants ε_1, ε_2, ε_3.[*] The second of these media is a medium of thickness $l_j = 2(j-1)a$ formed by the adsorbed layer and is assigned the dielectric properties of the bulk adsorbate. In both (i) and (ii) we have ignored "lateral" interactions which will be important in mobile monolayer adsorption (de Boer, 1968; Hill, 1946, 1947). This will be discussed later.

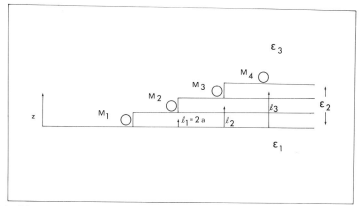

FIG. 6.1. Schematic representation of adsorbed layers. The molecules M_1, M_2, M_3 sit on filled layers which have the dielectric properties of the bulk adsorbate. [From Mahanty and Ninham (1974).]

[*] Strictly speaking, these assumptions would be incorrect if the adsorbed layers are of uniform thickness. In B. E. T. theory, however, the adsorbed molecules are assumed not to form uniform layers, but rather an assembly of piles, so that each adsorbed layer is partially exposed. In this situation the above assumption gives the energy of adsorption for each layer.

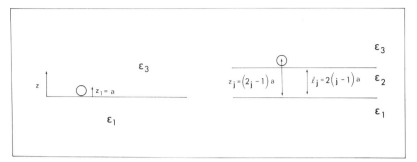

FIG. 6.2. Model for calculation of self-energies.

The self-energies of M_1 and M_j ($j > 1$) can be obtained from Chapter 4. First we recall from eqns. (4.63) and (4.68) that the dispersion self-energy (measured with respect to a zero which is the self-energy at infinite distance) for a molecule of radius a and polarizability $\alpha_3(\omega)$ at a distance z from the medium ε_1 can be written as*

$$E(z) \cong 2\hbar \int_0^\infty \alpha_3(i\xi)\, \text{Tr}\, \mathscr{G}_1(z, z)\, d\xi, \tag{6.2}$$

where

$$\text{Tr}\, \mathscr{G}_1(z, z) = -\frac{\Delta_{13}}{2\pi a^3}\left(\frac{\theta(z)}{\varepsilon_3} - \frac{\theta(-z)}{\varepsilon_1}\right)\left\{\frac{1}{\sqrt{\pi}}\exp\left(-\frac{z^2}{a^2}\right) - \frac{|z|}{a}\,\text{erfc}\left(\frac{|z|}{z}\right)\right.$$
$$\left. + \frac{a^3}{2\sqrt{\pi}|z|^3}\int_0^{|z|/a}\left[\exp\left(-t^2\right) - \exp\left(-\frac{z^2}{a^2}\right)\right]dt\right\}. \tag{6.3}$$

Note that as $z/a \to 0+$, we have

$$\text{Tr}\, \mathscr{G}_1(z, z) \to -\frac{\Delta_{13}}{2\pi a^3 \varepsilon_3}\left\{\frac{4}{3\sqrt{\pi}} - \left(\frac{z}{a}\right) + 0\left(\frac{z}{a}\right)^2\right\}, \tag{6.4}$$

while for $z > a$

$$\text{Tr}\, \mathscr{G}_1(z, z) \approx -\frac{\Delta_{13}}{2\pi a^3}\left\{2\left(\frac{a}{2z}\right)^3 - \frac{e^{-z^2/a^2}}{\sqrt{\pi}}\left(\frac{a}{z}\right)^4\left[1 + 0\left(\frac{a}{z}\right)^2\right]\right\}. \tag{6.5}$$

The structure of the Green function is such that the interaction energy is finite for very small distances $|z| \lesssim a$, and switches over to the large distance form, $E \propto 1/z^3$ already at a distance from the surface of only a

---

* For notational convenience here and subsequently we have factored out $\alpha(\omega)$ from the definition of $\mathscr{G}$ given in eqn. (4.17), with the form of $\alpha(\mathbf{r}, \omega)$ given by eqns. (4.20) and (4.21).

FIG. 6.3. An alternative model. Here the adsorbed molecule nestles closely into the bulk adsorbate, or the adsorbed film.

single atomic diameter. Taking $z = a$ in eqn. (6.3), we have explicitly for a molecule M_1 an adsorption energy of

$$E_1(a) = -\frac{\hbar c_1}{\pi a^3} \int_0^\infty d\xi \alpha_3(i\xi) \frac{\Delta_{13}}{\varepsilon_3}, \tag{6.6}$$

where c_1 is a numerical constant given by

$$c_1 = \left(-\tfrac{5}{4} \operatorname{erfc}(1) + \frac{e^{-1}}{2\sqrt{\pi}} + \frac{1}{4}\right) \approx 0 \cdot 16. \tag{6.7}$$

If we were to consider the alternative model in Fig. 6.3, the corresponding adsorption energy would be obtained by taking $z = 0$ in eqn. (4.70). The result is

$$E_1(0) = -\frac{2\hbar}{\pi^{1/2} a^3} \int_0^\infty \alpha_3(i\xi) \left(\frac{\varepsilon_1 - \varepsilon_3}{\varepsilon_1 \varepsilon_3}\right) \left(\frac{1}{2} + \frac{\Delta_{13}}{3}\right). \tag{6.8}$$

Equations (6.6) and (6.8) give the difference in energy $E_{ad} - E_g$ between a molecule in the adsorbed state and in the gas phase, and we have $Q_{int} = -(E_{ad} - E_g)$ where Q_{int} is the integral heat of adsorption. E.g. heats of adsorption for a homologous series of hydrocarbons have been calculated from self-energy by Chan and Richmond (1975), and agree with measured values. The difference in energy $E_1(0) - E_1(a)$ provides an estimate of the energy of activation which must be taken up by an adsorbed molecule in order to enable such a molecule to move over the surface. Using typical ultraviolet data for dielectrics, and carrying out the integrals over frequency in eqns. (6.6) and (6.8) one can show that $|\Delta E(a)| \lesssim |\tfrac{1}{2} E_1(0)|$ which agrees with intuitive guesses (de Boer, 1968) for the magnitude of the activation energy. Use of detailed ultraviolet spectral data for specific adsorption problems should provide reasonable estimates for this quantity. For the moment we pursue the first model, eqn. (6.6), for the adsorption energy.

The corresponding self-energy for a molecule M_j adsorbed into $(j-1)$ layers is given by

$$E_j = E[(2j-1)a] - E(\infty)$$
$$= 2\hbar \int_0^\infty \alpha_3(i\xi) \operatorname{Tr} \mathscr{G}_j[(2j-1)a] \, d\xi. \tag{6.9}$$

The Green function for this case is very much more complicated than $\mathcal{G}_1$, and can be obtained from eqn. (4.106). We make the approximation $\varepsilon_j(i\xi) = 1 + 4\pi N\alpha_j(i\xi)$, where N is the number density in medium j. Keeping only lowest order terms in polarizability we have, after a certain amount of algebra

$$\text{Tr}\,\mathcal{G}_j \approx \text{Tr}\,\mathcal{G}_1(a, a)[1 \to 2] + \text{Tr}\mathcal{G}_1[(2j-1)a][3 \to 2]. \qquad (6.10)$$

The notation $[1 \to 3]$ means that ε_1 is to be replaced by ε_3 everywhere in $\mathcal{G}_1$. The self-energy for molecule M_j consists then of two terms: (i) the energy of adsorption of a molecule of polarizability α_3 adsorbed from the gas phase 3 onto an infinite medium 2 together with (ii) the energy of interaction of a molecule 3 with medium 1 across a medium 2 of thickness $(2j-1)a$. This last term does not appear in the B.E.T. theory and is responsible for the $1/l^3$ isotherm for thick films. On the other hand the first term does not appear in the theories of Dzyaloshinskii, Lifshitz and Pitaevskii (1961), Halsey (1948), Cook (1948) and others. Using the results of eqns. (6.3) and (6.6) in eqn. (6.10) we have

$$E_j = -\frac{\hbar}{\pi a^3} \int_0^\infty d\xi \alpha_3(i\xi) \left[\frac{\Delta_{13}c_1}{\varepsilon_3} + \frac{\Delta_{12}}{4\varepsilon_2} \left\{ \frac{1}{(2j-1)^3} \right. \right.$$
$$\left. \left. - \frac{e^{-(2j-1)^2}}{\sqrt{\pi}(2j-1)^4} \left(1 + 0\left(\frac{1}{(2j-1)^2} \right) \right) \right\} \right]. \quad (6.11)$$

For all practical purposes, terms beyond the first in the second expansion can be dropped. If the alternative model of Fig. 6.3 is used the corresponding result is

$$E_j \approx -\frac{2\hbar}{\pi^{3/2}a^3} \int_0^\infty \alpha_3(i\xi)\, d\xi \left[\left(\frac{\varepsilon_2 - \varepsilon_3}{\varepsilon_2 \varepsilon_3} \right)\left(\frac{1}{2} + \frac{\Delta_{23}}{3} \right) + \frac{\sqrt{\pi}}{16}\frac{\Delta_{12}}{(2j-1)^3}\left(\frac{1}{\varepsilon_3} + \frac{1}{\varepsilon_2} \right) \right].$$

$$(6.12)$$

It should be noted that we are not considering the contribution to the self-energy due to short-range forces. In general we can expect this contribution to the energy of adsorption of a molecule in the first layer to be different from that of a molecule in subsequent layers. The net effect of taking account of short-range forces for all layers other than the first would be to shift the value of E_j, $j > 1$ by a constant, and the value of E_1 by (possibly) a different constant. The refinement will not affect the main conclusions, and hence we do not consider short-range forces.

6.3   The molecular partition function

The molecular function for M_j can now be constructed from eqns. (6.6) and (6.11). The fact that the adsorption energy of a molecule varies from layer to layer requires that the B.E.T. theory must be modified in the manner suggested by Cook (1948) and Dole (1948). However, for the moment we shall not include entropic effects associated with different configurations of the adsorbed layer which must be taken into account to make B.E.T. theory more realistic. In the notation of de Boer (1968) the equation for the isotherm can be written in the form

$$\sigma/\sigma_0 = \frac{\sum\limits_{j=1}^{\infty} j(n\tau_0/\sigma_0)^j \exp{(Q_j/k_BT)}}{1 + \sum\limits_{j=1}^{\infty} (n\tau_0/\sigma_0)^j \exp{(Q_j/k_BT)}}. \tag{6.13}$$

Here σ is the number of adsorbed molecules per unit area, σ_0 is the number of adsorption sites per unit area of surface of the adsorbent, and n is the number of adsorbate molecules which collide with unit area of the adsorbent per unit time. The time spent by a molecule on the j-th layer is τ_j, and the parameter τ_j is related to τ_0 through the relation

$$\tau_j = \tau_0 \exp{(Q_j/k_BT)}, \tag{6.14}$$

where Q_j, the heat of adsorption, is defined by

$$Q_j = -\sum_{j'=1}^{j} E_{j'}. \tag{6.15}$$

By comparison with §4.6 it can be shown that for large j the heat of adsorption consists of four terms: the contribution to the surface energy per particle of medium 2 at the interface [12]; the surface energy per particle of medium 2 at the interface [23]; the energy of interaction per particle of a molecule at the surface of medium 3 with medium 1 across a medium 2; the energy of condensation required to take j particles from the gas phase to the bulk liquid.

The relation between n and the properties of the adsorbate gas, assuming the gas to be nearly ideal, is

$$n = \frac{Np}{\sqrt{2\pi MRT}} \, \text{cm}^{-2}\,\text{s}^{-1}, \tag{6.16}$$

where p is the gas pressure in dyn cm^{-2}, N is Avogadro's constant, M the molecular weight and R the gas constant. Within the restrictions of the model eqn. (6.13) determines the adsorption isotherm. We proceed now to evaluation of the required sums.

From eqns. (6.15), (6.6) and (6.11) we have

$$Q_j/k_B T \approx K_1 + (j-1)K_2 + K_3 \sum_{j'=2}^{j} \frac{1}{(2j'-1)^3}, \tag{6.17}$$

where

$$K_1 = \frac{\hbar c_1}{k_B T \pi a^3} \int_0^\infty \alpha_3(i\xi) \frac{\Delta_{13}}{\varepsilon_3} d\xi; \quad K_2 = K_1(1 \to 2) \tag{6.18}$$

$$K_3 = \frac{4\hbar}{4 k_B T \pi a^3} \int_0^\infty \alpha_3(i\xi) \frac{\Delta_{12}}{\varepsilon_2} d\xi. \tag{6.19}$$

The sum required in eqn. (6.17) can be evaluated using the Euler–Maclaurin summation formula and has the asymptotic expression

$$\Sigma \equiv \sum_{j'=2}^{j} \frac{1}{(2j'-1)^3} \approx -1 + \frac{7}{8}\zeta(3) - \frac{1}{8}\left\{ \frac{1}{2(j+\frac{1}{2})^2} + \frac{1}{2(j+\frac{1}{2})^3} + 0\left(\frac{1}{j+\frac{1}{2}}\right)^4 \right\} \tag{6.20}$$

with $\zeta(3)$ a Riemann zeta function. Numerically we have

$$\Sigma = 0.052 - \frac{1}{16}\left\{ \frac{1}{j^2} + \frac{9}{4j^4} + 0\left(\frac{1}{j^5}\right) \right\}. \tag{6.21}$$

Keeping leading terms only of the last asymptotic expansion gives an approximation correct numerically to two decimal places even for $j = 2$, which will be sufficient for our purpose. Hence we can write

$$Q_j/k_B T = K_1, \qquad\qquad j = 1 \tag{6.22a}$$

$$Q_j/k_B T \sim (K_2 j + b - d/j^2), \qquad j = 2, 3, \ldots \tag{6.22b}$$

where

$$b = (K_1 - K_2 + 0.8d), \quad d = K_3/16. \tag{6.22c}$$

Then from eqn. (6.13) we have

$$\sigma/\sigma_0 \approx \frac{ze^{(K_1-K_2)} + e^b \sum_{j=2}^{\infty} jz^j e^{-d/j^2}}{1 + ze^{(K_1-K_2)} + e^b \sum_{j=2}^{\infty} z^j e^{-d/j^2}}, \tag{6.23}$$

where we have written $z = (n\tau_0/\sigma_0)e^{K_2}$. The parameter K_2 is in fact always positive, and the series converge as long as $z < 1$. When the series diverge

$\sigma/\sigma_0 \to \infty$, i.e., $p = p_0$, where p_0 is the saturated vapour pressure. Since $n \propto p$, the parameter z can be replaced by p/p_0.[*]

The remaining sums are quite complicated, and are carried out in Appendix C.

6.4 The adsorption isotherm

Thick films

Writing $z = e^{-B} = p/p_0$, we have from Appendix D

$$\sigma/\sigma_0 = \frac{(z e^{(K_1-K_2)} - z e^{(b-d)} + e^b S_2)}{(1 + z e^{(K_1-K_2)} - z e^{(b-d)} + e^b S_1)}, \qquad (6.24)$$

where

$$S_1 = \left(\frac{2^{4/3}\pi}{3}\right)\left(\frac{d}{B}\right)^{1/3} \frac{1}{(B^2 d)^{1/6}} e^{-(3/2)(B^2 d)^{1/3}}\left\{1 + 0\left(\frac{1}{(B^2 d)^{1/3}}\right)\right\} \qquad (6.25a)$$

$$S_2 = (d/B)^{1/3} S_1. \qquad (6.25b)$$

The approximation of eqn. (6.25a) holds only under the conditions

$$B \ll 1, \qquad d \gg 1, \qquad (B^2 d) \gg 1. \qquad (6.26)$$

The last inequality is equivalent to the condition that $p \ll p_0 e^{-1/\sqrt{d}}$. Then provided S_1 dominates both in numerator and denominator, we find

$$\sigma/\sigma_0 \approx (d/B)^{1/3} \qquad (6.27)$$

or substituting for d from eqn. (6.18) and (6.21), and writing $\sigma/\sigma_0 = l/2a$ where l is the thickness of the adsorbed film, we have

$$\ln(p/p_0) = -\frac{\hbar}{4\pi l^2 k_B T} \int_0^\infty \alpha_2(i\xi) \frac{\Delta_{12}}{\varepsilon_2} d\xi. \qquad (6.28)$$

[*] This identification is not quite correct. For imagine the special case that medium 1 is identical with medium 2, i.e., we consider the equilibrium of a liquid 2 with its vapour. In that case $d \equiv 0$ and $K_1 = K_2$. The series in eqn. (6.22) then diverge at the saturated vapour pressure: $z^0 = 1 = (n^0 \tau_0/\sigma_0) e^{K_2^0}$, where n^0 is the density of the gas corresponding to that of the saturated vapour pressure, and K_2^0 is given by eqn. (6.17), but with ε_3 evaluated at the density of the saturated vapour. Thus $z = p/p_0 e^{(K_2 - K_2^0)}$. The distinction will normally be unimportant when dispersion forces alone are involved, but will be significant if we deal with multimolecular adsorption from solution, or with adsorption of dipolar molecules. In these cases temperature-dependent (entropic) contributions to the van der Waals energies of adsorption and interaction and hence to the heats of adsorption can be as large as or larger in magnitude than the (energetic) ultraviolet contributions from the dispersion forces (Parsegian and Ninham, 1970).

This is to be compared with the Lifshitz expression based on continuum theory for thick films, which is (§3.5)

$$\ln (p/p_0) = -\frac{\hbar}{8\pi^2 l^3 k_B T n_2} \int_0^\infty \Delta_{12} \Delta_{23} \, d\xi. \qquad (6.29)$$

If in this formula we specialize to the case that dispersion forces only are involved, we can write for the gas phase $\varepsilon_3 = 1$, and for the adsorbed phase $\varepsilon_2 = 1 + 4\pi n_2 \alpha_2(i\xi)$ whence $\Delta_{23} \approx 2\pi n_2 \alpha_2(i\xi)$, and substituting into eqn. (6.24) recover an expression which coincides with our eqn. (6.28). The sole difference is that α_2 replaces α_3, but to the approximation considered, these are identical.

Thick films: influence of molecular size and B.E.T. theory

It can be shown that under the conditions of eqn. (6.26), the approximation which leads to the Lifshitz isotherm eqn. (6.28) is always valid. These conditions are equivalent to the statement

$$1 \gg \ln (p_0/p) \gg 1/\sqrt{d}. \qquad (6.30)$$

The first inequality implies that the pressure is not too low. The second is the more interesting. This condition cannot emerge from a continuum theory, since from eqn. (6.18) and (6.22c) $d \propto 1/a^3$ where a is the molecular radius, and the second inequality will then automatically be satisfied.

We now ask what happens when the sign of this inequality is reversed:

$$\ln (p_0/p) \ll 1/\sqrt{d}, \qquad d \gg 1, B \ll 1, B^2 d \ll 1. \qquad (6.31)$$

These conditions can be achieved if (1) the pressure becomes very close to the saturated vapour pressure; or (2) for a given pressure and fixed polarizability the size of the adsorbed molecules increases; or (3) if for a given pressure the parameter d, which is a measure of the dielectric properties of substrate and adsorbate, increases. Under the condition $\ln (p_0/p) \ll 1/\sqrt{d}$, $d \gg 1$, we have from Appendix D eqn. (D.9) and (D.10),

$$S_1 \approx (1/B)\{1 - \sqrt{\pi}/2(B^2 d)^{1/2} - (B^2 d) \ln B + 0(B^2 d)\} \qquad (6.32)$$

$$S_2 \approx (1/B^2)\{1 + \tfrac{1}{2}(B^2 d) \ln (B^2 d) + \cdots\}. \qquad (6.33)$$

Keeping only leading terms, we have from eqn. (6.23)

$$\left(\frac{\sigma}{\sigma_0}\right)_{(p \to p_0)} \approx \frac{k}{(p_0/p - 1)[1 - p/p_0 + k]}, \qquad k \approx \left\{\frac{e^b}{1 + e^{(K_1 - K_2)} - e^{(b-d)}}\right\} \qquad (6.34)$$

where we have written $B = \ln (p_0/p) \approx (p_0/p - 1)$. This isotherm is identical with that given by B.E.T. theory, an entirely natural result. Equivalently, for $p \to p_0$ we have

$$\ln (p/p_0) \approx 1/l. \tag{6.35}$$

The difference between B.E.T. theory and the present formulation is that the parameter k is determined and can be calculated without difficulty. (If the identification $z = \mathrm{e}^{-B} = (p/p_0)\mathrm{e}^{(K_1 - K_2^0)}$ is made there remains only one free parameter σ_0.)

Both isotherms, the Lifshitz form $\ln (p_0/p) \propto 1/l^3$ and B.E.T. form $\ln (p_0/p) \propto 1/l$ have been derived under the assumptions that $\ln (p_0/p) \ll 1$ and $d \gg 1$, and eqn. (6.35) especially uses quite drastic approximations. It is clear that in general depending on molecular size, pressure, and properties of the interacting species, a very wide range of behaviour can be expected, even for the special case of dispersion forces, and that attempts to force multimolecular adsorption curves to fit a form like $\ln (p_0/p) \propto 1/l^n$ may therefore be inappropriate. In general the complete expressions must be used. We can expect deviations from the Lifshitz isotherm at pressures $p \gtrsim p_0 \mathrm{e}^{-1/\sqrt{d}}$, although this criterion may be complicated by retardation in some cases. The B.E.T. theory proper is a special case of our formalism with $d = 0$.

Thin films: restricted adsorption

At low pressures $B = \ln (p_0/p) \gg 1$ and $z \ll 1$. If for fixed d we let $p \to 0$ in eqn. (6.23) we recover the Langmuir isotherm

$$\sigma/\sigma_0 \approx \frac{z\mathrm{e}^{(K_1 - K_2)}}{1 + z\mathrm{e}^{(K_1 - K_2)}} \equiv \frac{(n\tau_0/\sigma_0)\mathrm{e}^{Q_1/k_B T}}{1 + (n\tau_0/\sigma_0)\mathrm{e}^{Q_1/k_B T}}. \tag{6.36}$$

Again, however, we note that we now have a means of establishing connections between the polarizability and size of the adsorbate molecule and the parameters of the adsorption isotherm. A more interesting result emerges from eqn. (6.23) if we recognize that for most adsorption problems involving gases d is large and positive. Then writing out the sum which occurs in the numerator of eqn. (6.23) we have

$$\sum_{j=2}^{\infty} jz^j \mathrm{e}^{-d/j^2} = 2\mathrm{e}^{-(2B+d/4)} + 3\mathrm{e}^{-(3B+d/9)} + 5\mathrm{e}^{-(4B+d/16)} + \cdots. \tag{6.37}$$

Since B and d are both large and positive, we can expect one of the exponentials to dominate the sum. For example if $5d/36 > B > 7d/144$ the second term will be dominant and our isotherm would have the form

$$\sigma/\sigma_0 \approx \mathrm{e}^{(K_1 - K_2)} \frac{[\mathrm{e}^{-B} + 3\mathrm{e}^{0.8d}\mathrm{e}^{-(3B+d/9)}]}{1 + z\mathrm{e}^{(K_1 - K_2)}[\mathrm{e}^{-B} + \mathrm{e}^{0.8d}\mathrm{e}^{-(3B+d/9)}]} \approx 3. \tag{6.38}$$

As the pressure increases through the range $5d/36 > B > 7d/144$, σ/σ_0 will remain substantially constant at the value 3, and then jump to the value 4. This step-wise behaviour of the isotherm does not necessarily follow, and again depends on polarizability, size, and dielectric properties in a predictable manner. Step-wise behaviour of plots of σ/σ_0 against p/p_0 need not indicate composite surfaces.

One additional result which emerges from eqn. (6.23) is that when d increases, i.e., if the molecular size increases at fixed values of other parameters, monolayer adsorption is favoured over multilayer adsorption.

6.5 Lateral interactions in monolayer adsorption

The above analysis is based on an extension of B.E.T. theory which ignores the effect of lateral interactions between the molecules in adsorbed layers. Lateral interactions between the molecules once the first layer is complete are relatively unimportant. They can be taken into account, for example, using the alternative model illustrated in Fig. 6.3. The adsorption energy clearly differs very little from that given by the model of Fig. 6.2. The effect of dispersion forces between adsorbate molecules in a monolayer has been extensively treated by Hill (1946, 1947). We shall not repeat his analysis, but remark that the effective interaction energy is not that given by Hill, who assumed a value appropriate to two adsorbate molecules in a vacuum. The presence of a substrate alters the interaction energy from this value (Mahanty and Ninham, 1973; Richmond and Sarkies, 1972; Duniec and Ninham, 1976). A simple method of incorporating this effect into our formalism is simply to take for the value of the energy of adsorption a value appropriate to the degree of coverage. The corresponding dispersion self-energy is that of a molecule M_1 in a layer of thickness $2a$ and immersed in a medium of dielectric properties $\varepsilon_2' = (1 - \sigma/\sigma_0)\varepsilon_3 + \sigma/\sigma_0\varepsilon_2$. Use of this interpolation in our equations will reproduce a result for monolayer adsorption equivalent to Hill's treatment. To see this, we proceed as follows. For the situation depicted in Fig. 6.4, the energy of adsorption is,

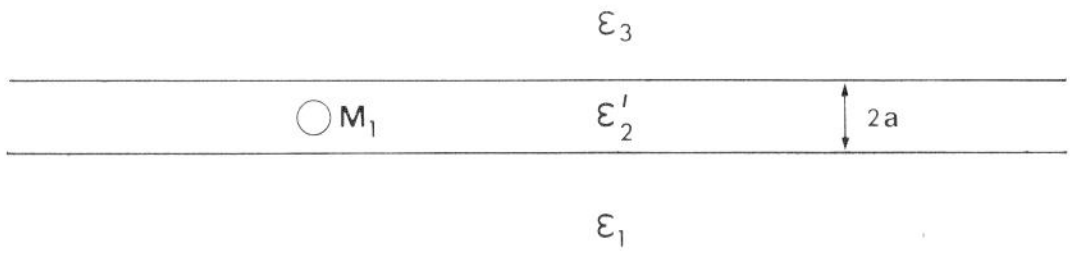

FIG. 6.4. Model for dealing with lateral interactions. Here when a fraction σ/σ_0 of the surface is covered the first layer is assigned dielectric properties which are a weighted sum of those of the bulk adsorbate and gas phase: $\varepsilon_2' = (1 - \sigma/\sigma_0)\varepsilon_3 + \sigma/\sigma_0\varepsilon_2$.

using Appendix A and eqn. (4.106)

$$E_1 = -\frac{2\hbar}{\pi^{3/2}a^3}\int_0^\infty \alpha_3 \frac{(\varepsilon_2'-\varepsilon_3)}{(\varepsilon_2'\varepsilon_3)}\,d\xi - \frac{\hbar c_1}{\pi a^3}\int_0^\infty \alpha_3 \left(\frac{\Delta_{12}'}{\varepsilon_2'}+\frac{\Delta_{32}'}{\varepsilon_2'}\right)d\xi. \quad (6.39)$$

The first term is the difference in self-energy per particle between one adsorbed into a monolayer of dielectric properties $\varepsilon_2' = 1 + \sigma/\sigma_0(\varepsilon_2 - \varepsilon_3)$ and therefore includes molecular interactions if these are treated as unmodified by the presence of substrate or the gas phase. The second term measures these effects. For an unfilled monolayer, entropic effects must also be included in the heat of adsorption. The inclusion of these effects involves a subtle thermodynamic argument (de Boer, 1968), the net result of which is that the heat of adsorption is to be taken as

$$Q_1 = T\,\Delta S - E_1 = -\frac{k_B T\theta}{1-\theta} - E_1. \quad (6.40)$$

From eqn. (6.23) we then obtain for small values of p/p_0

$$\theta = \sigma/\sigma_0 = \frac{(p/p_0)e^{Q_1/k_B T}}{1 + (p/p_0)e^{Q_1/k_B T}} \quad (6.41)$$

or

$$p/p_0 = \frac{\theta}{1-\theta}\,e^{(\theta/1-\theta)}e^{E_1/k_B T}. \quad (6.42)$$

To recover the Hill isotherm it is simply necessary to drop the second term of eqn. (6.39), i.e., to ignore the influence of substrate on molecular interactions, and in the remaining term

$$E_1 \approx -\frac{2\hbar}{\pi^{3/2}a^3}\int_0^\infty \alpha_3 \frac{(\varepsilon_2'-\varepsilon_3)}{\varepsilon_2'\varepsilon_3}\,d\xi \quad (6.43)$$

to (1) drop the denominator in $\varepsilon_2'\varepsilon_3$, (2) put $\varepsilon_3 \approx 1$ (the gas phase has dielectric properties close to those of a vacuum), (3) assume $\varepsilon_2 = 1 + 4\pi n_2\alpha_3$ (Lorentz form), where n_2 is the density of the liquid phase, and (4) to substitute $\varepsilon_2' - \varepsilon_3 = 1 + \theta(\varepsilon_2 - \varepsilon_3)$. This gives

$$p/p_0 = (\theta/1-\theta)e^{(\theta/1-\theta)}e^{-\gamma\theta/k_B T}, \quad (6.44)$$

where

$$\gamma = \frac{8\hbar\pi}{\pi^{3/2}a^3}n_2\int_0^\infty \alpha_3^2\,d\xi. \quad (6.45)$$

If we use the formula for oscillator dispersion at a single frequency

$$\alpha_3 = \left(\frac{\alpha_3^0}{1-\omega^2/\omega_0^2}\right) \quad (6.46)$$

and interpret the radius of a molecule in terms of our model, we find

$$\gamma \cong (2a_2\sigma_0), \tag{6.47}$$

where a_2 is the usual parameter which occurs in the two-dimensional van der Waals equation of state. Then eqn. (6.44) and the Hill isotherm coincide.

In general, however, we see from eqn. (6.39) that the manner of "two-dimensional" condensation depends on the nature of the substrate, which must affect the strength of lateral interactions.

The effects of temperature and retardation can easily be built into our formalism in principle as can those due to spatial dispersion. The first effect will normally be unimportant when dispersion forces alone are involved, but will be of importance in adsorption from aqueous solution.

6.6 Phase transitions in polymer solutions

A further application where the concept of dispersion self-energy can be put to good use is in the theory of polymer solutions. (For a review of literature on the subject of polymer conformations and polymer adsorption from the theoretical standpoint up to 1970 see Barber and Ninham (1970).) A great deal of work has been devoted to the study of the mean square radius and other properties of model polymers. Mathematically, the problem has been formulated in such a way that it entails the solution of a self-avoiding random walk, a problem of immense intractability. Especially as a result of Edwards' work (Edwards, 1965) it seems now generally agreed by theorists, and confirmed by experiments, that the original prediction of Flory (1949) that $\langle R^2 \rangle \approx N^{6/5}$ is correct. Here $\langle R^2 \rangle$ is the mean square radius of the polymer, accessible from light scattering experiments, and N is the number of links in the polymer chain. Experimentally, for some polymer-solvent solutions, it is found that at a certain temperature, known as the Flory or theta temperature, the dependence of $\langle R^2 \rangle$ on N changes to the form $\langle R^2 \rangle \sim N$, which is that predicted for a volumeless random flight configuration. The effect of the self-avoidance condition, or excluded volume, is to expand the polymer relative to the random flight configuration. If polymer–polymer interactions are favoured over polymer–solvent interactions the polymer tends to contract. Consequently, in a poor solvent it is possible for a contraction due to the solvent interactions with the polymer to exactly compensate the expansion due to excluded volume. While a large literature has been devoted to models based on the usual virial expansion methods of statistical mechanics, and to lattice walk models, the transition from $\langle R^2 \rangle \sim N^{6/5}$ to $\langle R^2 \rangle \sim N$ cannot be predicted by these approaches, and all theories contain undetermined parameters which restrict their application to real polymer systems. The theory developed below follows Chan and Ninham (1974) who use the self-energy concept together with the self-consistent field analysis of Edwards (1965) as simplified by de Gennes (1969) to relate such parameters to directly measurable characteristics of a polymer and solutions.

In de Gennes' (1969) formulation of the problem, the polymer is pictured as a collection of beads located at positions $\mathbf{r}_1, \mathbf{r}_2, \ldots, \mathbf{r}_N$ held at a fixed distance $b = |\mathbf{r}_{n+1} - \mathbf{r}_n|$ apart. In the presence of an external force field, the mean orientation of the $(n, n+1)$ link is $\langle (\mathbf{r}_{n+1} - \mathbf{r}_n) \rangle \equiv \mathbf{u}_n$. For small orientation distortion of the link $|\mathbf{u}_n| \ll b$, $u_n = 0$ corresponding to complete orientational disorder, the associated decrease in entropy of the link* is

$$\Delta S_n = -\frac{3 k_B}{2 b^2} u_n^2 \tag{6.48}$$

which amounts to a change in the free energy of

$$\Delta F_n = -T \Delta S_n = \frac{3 k_B T}{2 b^2} u_n^2. \tag{6.49}$$

The total free energy change

$$F = \frac{3 k_B T}{2 b^2} \sum_n (r_{n+1} - r_n)^2 \rightarrow \frac{3 k_B T}{2 b^2} \left(\frac{\partial \mathbf{r}}{\partial n}\right)^2 \tag{6.50}$$

can then be obtained by summing over all links. Differentiating the free energy with respect to $\mathbf{r}_n$ yields the force acting on the n-th bead. In the

----

* This can be seen as follows: If $p(\theta)$ denotes the probability that the link has orientation θ, measured with respect to $\mathbf{u}$, the most general form of p which leads to an average polarization $\mathbf{u}$ is

$$p = \tfrac{1}{2}(1 + 3u \cos \theta + u^2 f(\theta) + \cdots)$$

with f arbitrary except that the normalization of p,

$$\int_0^\pi p(\theta) \sin \theta \, d\theta = 1,$$

implies that

$$\int_0^\pi f(\theta) \sin \theta \, d\theta = 0.$$

The entropy of the link is

$$S = -k_B \int_0^\pi p \ln p \sin \theta \, d\theta$$

which leads directly to (6.48).

continuum limit, this force is

$$\mathcal{F}(\mathbf{r}) = -\frac{\partial F}{\partial \mathbf{r}_n} = \frac{3k_BT}{b^2}\frac{\partial^2 \mathbf{r}}{\partial n^2}. \tag{6.51}$$

If the force can be derived from a potential $\mathcal{F} = -\nabla\phi$, this expression may be integrated to give

$$\phi(\mathbf{r}) - \frac{3k_BT}{2b^2}\left(\frac{\partial \mathbf{r}}{\partial n}\right)^2 = \text{constant}. \tag{6.52}$$

In dealing with the excluded volume problem, de Gennes (1969) shows further that $\phi(\mathbf{r})$ is proportional to $\tilde{\rho}(\mathbf{r})$ the average segment or bead density at $\mathbf{r}$. Further, if one assumes radial symmetry, the number of segments in the ranges $\mathbf{r}$ to $\mathbf{r} + \mathbf{dr}$ is

$$dn = 4\pi r^2 \tilde{\rho}(r)\, dr. \tag{6.53}$$

The average bead density $\tilde{\rho}(r)$ so defined is such that N, the total number of polymer links, is given by

$$N = \int_0^{R_0} 4\pi r^2 \tilde{\rho}(r)\, dr, \tag{6.54}$$

where R_0 is the root mean square radius. Hence by setting $\phi = k_BTv\tilde{\rho}$, where v is the "excluded volume parameter," $\tilde{\rho}(r)$ may be determined self-consistently. If the excluded volume parameter v is constant, substitution of eqns. (6.53) and of $\phi = k_BTv\tilde{\rho}$ into (6.52) gives $\tilde{\rho}(r) \sim 1/r^{4/3}$; and from (6.54) $\langle R^2 \rangle \sim N^{6/5}$, and no theta point exists.

Conformation change via self-energy

To proceed further for real polymers, we require an expression not for v, but for the free-energy of a link as an equivalent one-body potential of a polymer segment. This potential must be density and temperature-dependent, and should be expressed in terms of the dielectric properties of the polymer solution. In principle, knowledge of the dielectric properties of solvent and polymer should provide implicitly a good description of the polymer–polymer, polymer–solvent and solvent–solvent interactions which are ultimately responsible for the chain configuration.

We confine our attention to non-aqueous solvents, where dispersion forces are the dominant interaction mechanism. Then with each polymer segment we can associate a one-body potential which is its dispersion self-energy. If the polarizability of each segment of the chain is assumed isotropic and each segment has a radius a, the self-energy per link is from eqn. (4.24) (generalized to include effects of temperature)

$$F_{\!\!/\!} = \frac{4\pi}{a^3\pi^{3/2}}\, k_B T \sum_{n=0}^{\infty}{}' \frac{\alpha_{\!/\!}(i\xi_n)}{\varepsilon_M(i\xi_n)}, \tag{6.55}$$

where ε_M is the dielectric susceptibility of the solution. $F_{\!/\!}$ provides a measure of the interaction between the polymer segment and the solvent as well as the interaction between the polymer segment and its neighbours. To be precise, one should also take into account the spatial distribution of the polymer segments, that is spatial variations in ε_M, when evaluating the self-energy.* However only molecules in the neighbourhood of the polymer segment in question contribute significantly to its self-energy, so that for a dilute polymer solution, a condition usually satisfied in practice, we can account for the distribution of other polymer segments by writing

$$\varepsilon_M = \varepsilon_{\!\!d} + \tilde{\hbar}\, \frac{\partial \varepsilon}{\partial \tilde{\hbar}}, \tag{6.56}$$

where $\varepsilon_{\!\!d}$ is the dielectric susceptibility of the pure solvent, $\tilde{\hbar}$ the local density of polymer segments, and the correction term to $\varepsilon_{\!\!d}$ is always small.

Returning now to eqn. (6.52) we write

$$\phi(\mathbf{r}) = \frac{4\pi}{a^3\pi^{3/2}}\, k_B T \left\{ \sum_n{}' \frac{\alpha_{\!/\!}(i\xi_n)}{\varepsilon_{\!\!d} + \tilde{\hbar}\, \dfrac{\partial \varepsilon_M}{\partial \tilde{\hbar}}} - \sum_n{}' \frac{\alpha_{\!/\!}(i\xi_n)}{\varepsilon_{\!\!d}} \right\}, \tag{6.57}$$

---

*This can be seen as follows: The dispersion self-energy of a molecule given by eqn. (4.24) is derived from a Green function $\mathcal{G}(r, r'; \omega)$ which for a homogeneous medium satisfies the equation

$$\varepsilon_M \nabla^2 \mathcal{G} = \delta(\mathbf{r} - \mathbf{r}').$$

If the dielectric susceptibility of the medium ε_M is a function of the coordinates (e.g., due to a spatial distribution of polymer segments) the equation for $\mathcal{G}$ should be

$$\varepsilon_M \nabla^2 \mathcal{G} + \nabla \varepsilon_M . \nabla \mathcal{G} = \delta(\mathbf{r} - \mathbf{r}').$$

Provided $\varepsilon_{\!\!d} \gg \tilde{\hbar}(\partial \varepsilon_M / \partial \tilde{\hbar})$ and the problem possesses radial symmetry, we can write

$$\varepsilon_M \frac{\partial^2}{\partial r^2}\mathcal{G} + \left(\frac{2\varepsilon_M}{r} + \frac{\partial \varepsilon_M}{\partial \tilde{\hbar}}\frac{\partial \tilde{\hbar}}{\partial r} \right) \frac{\partial \mathcal{G}}{\partial r} = \delta(\mathbf{r} - \mathbf{r}').$$

For a dilute polymer solution $\tilde{\hbar}(r)$ is just the local polymer density. When the solution is near the theta point $\tilde{\hbar}(r) \propto 1/r$. Using this value of $\tilde{\hbar}(r)$ above, we get

$$\varepsilon_M \frac{\partial^2}{\partial r^2}\mathcal{G} + \frac{2\varepsilon}{r}\left(1 - \frac{\tilde{\hbar}}{2\varepsilon_M}\frac{\partial \varepsilon_M}{\partial \tilde{\hbar}} \right) \frac{\partial \mathcal{G}}{\partial r} = \delta(\mathbf{r} - \mathbf{r}')$$

and since $\varepsilon_M \gg \tilde{\hbar}(\partial \varepsilon_M / \partial \tilde{\hbar})$ we can neglect spatial variations in ε_M.

where the second term describes the segment–polymer interaction in the absence of all other polymer segments. With this choice of the zero of potential, it follows that the constant of integration in (6.52) is zero. We now seek a solution of (6.52) for which $\tilde{\hbar}$ is radially symmetric. Expanding the denominator of (6.57) in powers of $\tilde{\hbar}$ and substituting into (6.52) we have

$$
-\frac{4\pi}{a^3\pi^{3/2}}k_BT\left\{\sum_n{}'\frac{\alpha_{\hbar}}{\varepsilon_{\vartheta}^2}\frac{\partial\varepsilon}{\partial\tilde{\hbar}}\tilde{\hbar}^3(r)-\sum_n{}'\frac{\alpha_{\hbar}}{\varepsilon_{\vartheta}^3}\left(\frac{\partial\varepsilon}{\partial\tilde{\hbar}}\right)^2\tilde{\hbar}^4(r)\right.
$$

$$
\left.+0(\tilde{\hbar}^5)\right\}=\frac{3k_BT}{2b^2}\frac{1}{(4\pi r^2)^2}. \qquad (6.58)
$$

In general the $\tilde{\hbar}^3$ term always dominates the left hand side and we have

$$
\tilde{\hbar}(r)\approx r^{-4/3}; \qquad \langle R^2\rangle\sim N^{6/5} \qquad (6.59)
$$

which is the familiar situation where excluded volume effects dominate. On the other hand, when the coefficient of $\tilde{\hbar}^3$ becomes identically zero, then

$$
\tilde{\hbar}(r)\approx r^{-1}, \qquad \langle R^2\rangle\approx N \qquad (6.60)
$$

and the polymer behaves like a chain in random flight. The singular situation when the coefficients of $\tilde{\hbar}^3(r)$ vanishes corresponds to the theta point. The physical significance of the coefficients of $\tilde{\hbar}(r)$ in (6.58) can be seen as follows. For non-polar materials we have $\varepsilon=1+4\pi n\alpha$. For a dilute polymer solution we have correspondingly

$$
\varepsilon_M=1+4\pi\tilde{\vartheta}\alpha_{\vartheta}+4\pi\tilde{\hbar}\alpha_{\hbar} \qquad (6.61)
$$

with the constraint $\tilde{\vartheta}+\tilde{\hbar}\approx n_{\vartheta}$, where α_{ϑ} is the polarizability of a solvent molecule, $\tilde{\vartheta}$ and n_{ϑ} are densities of solvent molecules in the solution and pure solvent respectively. Thus

$$
\frac{\partial\varepsilon}{\partial\tilde{\hbar}}\approx4\pi(\alpha_{\hbar}-\alpha_{\vartheta}). \qquad (6.62)
$$

Inserting this result into (6.58) we have for the coefficient of $\tilde{\hbar}^3(r)$

$$
a_3=-\frac{(4\pi)^2}{a^3\pi^{3/2}}k_BT\sum_n{}'\frac{\alpha_{\hbar}\alpha_{\hbar}-\alpha_{\hbar}\alpha_{\vartheta}}{\varepsilon_{\vartheta}^2} \qquad (6.63)
$$

and for the coefficient of $\tilde{\hbar}^4(r)$

$$
a_4=\frac{(4\pi)^3}{a^3\pi^{3/2}}k_BT\sum_n{}'\frac{(\alpha_{\hbar}\alpha_{\hbar}\alpha_{\hbar}+\alpha_{\hbar}\alpha_{\vartheta}\alpha_{\vartheta}-2\alpha_{\hbar}\alpha_{\hbar}\alpha_{\vartheta})}{\varepsilon_{\vartheta}^3}. \qquad (6.64)
$$

It is clear that a_4 describes the competition between the two-body solvent–polymer and polymer–polymer interactions and a_4 describes the competition among the three-body polymer–polymer–polymer, polymer–polymer–solvent and polymer–solvent–solvent interactions. In general two-body interactions are dominant and they give rise to the familiar $\langle R^2 \rangle \approx N^{6/5}$ behaviour. However, under special circumstances when the two-body interactions cancel exactly the three-body terms then give $\langle R^2 \rangle \approx N$, the random flight behaviour. The condition under which the two-body terms cancel (theta point) depends on a delicate balance of the temperature-dependent and other dispersion forces in the complicated sum over imaginary frequencies. The existence of a theta point for a given polymer/solvent system depends critically on the dielectric properties of both solvent and polymer as well as on the temperature.

Application to real polymer solutions

The assumptions made so far are: (1) The polymer has been modelled as a string of beads a distance b apart, each link of which has an isotropic polarizability. (2) The solvent surrounding each link is assumed isotropic and structureless. (3) Hindered rotation is not taken into account. (4) Short-range (hard core) forces are not considered, but are unimportant near the theta point. Evidently these assumptions are fairly extreme. Nonetheless the order of magnitude agreement of the theory with observation does emerge even at this level of sophistication (Chan and Ninham, 1974). For most systems of interest the required dielectric properties can be summarized by the simple forms

$$\varepsilon_j = 1 + 4\pi n_i \alpha_i; \qquad j = \text{\it s}, \text{\it p} \text{ for solvent or polymer} \qquad (6.65a)$$

$$\varepsilon(i\xi) = 1 + \frac{(n^2 - 1)}{1 + \xi^2/\omega_{uv}^2}; \qquad n = \text{refractive index}. \qquad (6.65b)$$

The value of n_i, the density of molecules which contribute to dielectric dispersion in a pure substance is uncertain, but it seems reasonable to assume that its value for the solvent or polymer do not differ greatly. The coefficient of $\bar{h}^3(r)$ in (6.63) can then be written as

$$a_3 = \frac{1}{a^3 \pi^{3/2}} \frac{k_B T}{n_0^2} {\sum_n}' \frac{(\varepsilon_p - 1)}{\varepsilon_s^2} \{(\varepsilon_p - 1) - (\varepsilon_s - 1)\}, \qquad (6.66)$$

where $n_s \approx n_p \approx n_0$. For a number of polymer/solvent systems this expression can be evaluated as a function of temperature (ε_p and ε_s both depend on temperature). Detailed calculations show that predicted theta temperatures are in reasonable agreement with experiment. The necessary condition that a theta point exist deserves note. We examine the two terms in braces in (6.66). In Fig. 6.5 each term, as a function of imaginary frequency is represented schematically, where $\hbar\omega_{uv}^p$ and $\hbar\omega_{uv}^s$ correspond

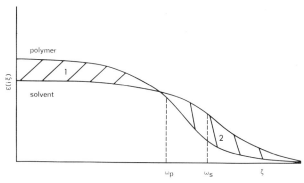

FIG. 6.5. Schematic representation of the dielectric susceptibility as a function of imaginary frequency for polymer/solvent systems. [From Chan and Ninham (1974).]

to first ionization potentials of polymer I_p and solvent I_s respectively. The summation over frequencies in (6.66) samples the difference between the two curves at intervals $\varepsilon_n = 2\pi n k_B T/\hbar$. Geometrically, this summation is roughly the same as the difference between the two shaded areas labelled 1 and 2. So for a theta point to exist, a delicate balance of the absorption spectra of the polymer and solvent is needed. From this we conclude that for a given polymer we have the general rule $\varepsilon_p(0) > \varepsilon_s(0)$; $I_p < I_s$, or $\varepsilon_p(0) < \varepsilon_s(0)$; $I_p > I_s$ for the static dielectric constant and ionization potential of the theta solvent. This rule does indeed appear to be satisfied by the polymer/solvent systems listed in Table 6.1. The rule stated above is expected to be satisfied by the non-polar polymers in non-polar solvents. When the static dielectric constants of the polymer and solvent are more nearly equal, a more detailed knowledge of the higher ionization potentials and corresponding oscillator strengths may be necessary to determine the theta point. In general it is possible for the curves of Fig. 6.5 to cross several times, in a manner precisely analogous to that discussed in §3.5. As for cases where the polymer segments and/or the solvent

Table 6.1. Data on polymers (actually monomer data) and solvents

System	$\Theta\,°K$	First ionization potential (eV)	Refractive index
Polystyrene in	304	8·47	1·55
decalin		9·61	1·48
Polyisobutene in	249	9·23	1·49
ethyl benzene		8·76	1·55
Polyethylene in	434	9·90	1·44
diphenyl ether		8·82	1·51

molecules possess permanent dipole moments (e.g., water) the position
and degree of Debye relaxations at lower frequencies will also influence
the theta point. If systems involving polyelectrolytes are concerned, the
analysis must be supplemented to include an electrostatic contribution to
the potential ϕ. The effect of charged segments in ionic solution is to
expand the polymer, as has been discussed by Richmond (1973).

The conformation of an adsorbed polymer, including solvent–polymer
and polymer–substrate interactions is still a completely open question, the
problem itself being obviously of key importance both for its applications
to colloid science and to biology. If polymer–solvent interactions are
ignored however, a complete theory has been given (Chan, Mitchell,
Ninham and White, 1975).

6.7 Some remarks on the consequences of inhomogeneity at surfaces and on adsorption

Until now we have assumed that bodies exist in well-defined geometric
shapes, and that interface profiles span a distance of the order of an
interatomic spacing. While numerical calculations of the surface energies
of pure hydrocarbon liquids and other systems referred to in Chapter 4
indicate that interface widths are indeed often of the order of interatomic
distances, for many applications this will certainly not be so. For example,
if we think of a surface covered with an adsorbed polymer, it is clear that
to characterize the adsorbed coating as a layer of given thickness and
uniform dielectric properties would generally be incorrect. Parsegian and
Weiss (1972) have studied by van Kampen's method the interaction
between two semi-infinite homogeneous bodies across a planar slab. In
their model there is a transition layer on either side of the slab where the
dielectric susceptibility varies exponentially in the direction perpendicular
to the slab (Fig. 6.6). The algebra is straightforward involving the solution
of

$$\nabla(\varepsilon\phi) = \nabla^2\phi + \frac{\partial \ln \varepsilon}{\partial z}\frac{\partial \phi}{\partial z} = 0 \qquad (6.67)$$

with the usual boundary conditions. An exponential profile for the
variation of ε in the transition layers was chosen for ease of solution, but
analytic solutions for the interaction energy obtainable for other assumed
profiles. The results are what one would have expected. When the
distance l between the slabs is much greater than the thickness d of the
transition region, the system behaves as two semi-infinite media ε_1 across
ε_3. On the other hand, if $d \gtrsim l$ the system behaves in a manner very
similar to a triple film. Unfortunately, except to underline the importance

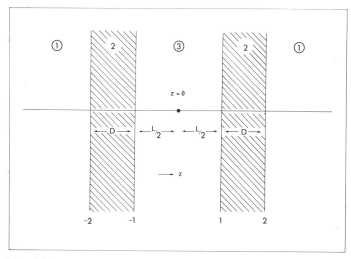

FIG. 6.6. Model for interaction between bodies covered with inhomogeneous layers. In media 1 and 3 the dielectric properties are constant, but in the layers 2 vary smoothly from those of material 1 to those of material 3, with a profile $\varepsilon_2(z) = \Gamma e^{-\gamma|z|}$. [From Parsegian and Weiss (1972).]

of inhomogeneity in interactions between colloidal bodies, such model calculations have little quantitative value. Apart from the fact that the interaction energy will depend critically on the assumed profile for variation of the dielectric properties of the adsorbed coat (which is not known), it seems obvious that the shape of the profile will change as the distance between the bodies decreases. Even if the density profile at a single surface is known, the question of interaction energies must include consideration of bulk and surface energies in a self-consistent manner. Langbein (1974) has also studied the interaction between two spheres whose susceptibility varies continuously as a function of the radial coordinates, and his results applied to the interaction of spheres coated with non-uniform adsorbed polymer by Kiefer, Parsegian and Weiss (1975).

Some other examples of problems in colloid science where a fixed step-like variation of dielectric susceptibility will not always occur in real systems may be worth pointing out. Consider a single solid or liquid body in equilibrium with a gas or solution. Suppose that the gas or solute molecules are adsorbed, and for simplicity of argument, that the conditions for multilayer adsorption hold. If this single body with its adsorbed coat now comes close to a similar body, the conditions for thermodynamic equilibrium determining the thickness of the coats will be altered. The full interaction problem must include contributions to the free energy of the system due to changes in thickness and density of the adsorbed coats.

Such considerations are essential to questions involving, for instance, the effect of adsorption on the stability of aerosols of solid bodies, of colloidal dispersions of solids in liquids, and of the swelling and shrinking of porous bodies caused by adsorption. The thermodynamics of these processes has been developed by Ash, Everett and Radke (1973), and a statistical mechanical theory can be developed by the methods of §6.1. With adsorbed polymers a similar process of rearrangement of density profile, and the formation of polymer "bridges" will occur due to the interactions. The last is a problem in questions concerning adhesion and lubrication, flocculation of soils by polymers, and in many biological problems. As mentioned already, Parsegian and Gingell (1973) have carried out a number of model calculations on biological cell–cell and cell–substrate interactions. Making what appear to be quite reasonable assumptions about the dielectric and charge properties of the cell surface and its vicinity, they showed that the DLVO theory of colloid stability, i.e., a balance between electrostatic and dispersion forces, is sufficient to give forces between cells large enough to hold them together in tissues. Their calculations are by no means simple-minded; they include the effect of charge regulation (Ninham and Parsegian, 1971) and (in revised form) the complications introduced by electrolyte solutions. The forces can be quite specific, and it appears that it may be unnecessary to invoke additional mechanisms like calcium binding or polymer bridging to explain cell adhesion. Their conclusions are probably correct, but the complications alluded to above have to be constantly kept in mind in extrapolating from the relatively well-defined area of colloid science to biology. Model calculations based on the theory of adsorbed films of Dzyaloshinskii, Lifshitz and Pitaevskii (1961) for the interaction of coated cell membranes have been made by Ninham and Richmond (1973). The theory confirms that adsorbed coats in equilibrium with an intervening solution can act as a "glue," or will thicken locally, or thin to zero thickness in the presence of a foreign body.

References

Ash, S. G., Everett, D. H. and Radke, C. (1973). *J. Chem. Soc., Farad. II* **69**, 1256.

Barber, M. N. and Ninham, B. W. (1970). "Random and Restricted Walks," Gordon and Breach, New York.

Brunauer, S., Emmett, P. H. and Teller, E. (1938). *J. Amer. Chem. Soc.* **60**, 309.

Brunauer, S., Copeland, L. E. and Kantro, D. L. (1967). "The Solid-Gas Interface" (ed. E. A. Flood), Marcel Dekker, New York, p. 77.

Chan, D. and Ninham, B. W. (1974). *J. Chem. Soc., Farad. II* **70**, 586.

Chan, D., Mitchell, D. J., Ninham, B. W. and White, L. (1975). *J. Chem. Soc., Farad. II* **71**, 235.

Chan, D. and Richmond, D. (1975). *Disc. Farad. Soc.* **59**, 13.

Cook, M. A. (1948). *J. Amer. Chem. Soc.* **70**, 2925.

de Boer, J. H. (1968). "The Dynamical Character of Adsorption," Oxford U.P., London.

de Gennes, P. G. (1969). *Rep. Progr. Phys.* **32**, 187.

Dole, M. (1948). *J. Chem. Phys.* **16**, 25.

Duniec, J. T. and Ninham, B. W. (1976). *J. Chem. Soc., Farad. II* (to appear).

Dzyaloshinskii, I. E., Lifshitz, E. M. and Pitaevskii, L. P. (1961). *Adv. Phys.* **10**, 165.

Edwards, S. F. (1965). *Proc. Phys. Soc.* **85**, 613.

Erdelyi, A., Magnus, W., Oberhettinger, F. and Tricomi, F. G. (1953). "Higher Transcendental Functions," Bateman Manuscript Project, McGraw-Hill, New York. Vol. 1, p. 27.

Flory, P. J. (1949). *J. Chem. Phys.* **17**, 303.

Halsey, G. (1948). *J. Chem. Phys.* **16**, 931.

Hill, T. L. (1946). *J. Chem. Phys.* **14**, 441.

Hill, T. L. (1947). *J. Chem. Phys.* **15**, 767.

Kiefer, J. E., Parsegian, V. A. and Weiss, G. H. (1975). *J. Coll. Interface Sci.* **51**, 543.

Langbein, D. (1974). "Theory of van der Waals Attraction," §48. Springer Tracts in Modern Physics, Springer-Verlag, Berlin. **72**, 139 pp.

Landau, L. D. and Lifshitz, E. M. (1958). "Statistical Physics," §146. Pergamon, London.

Mahanty, J. and Ninham, B. W. (1973). *J. Phys. A* **6**, 1140.

Mahanty, J. and Ninham, B. W. (1974). *J. Chem. Soc., Farad. II* **70**, 637,

Mahanty, J. and Ninham, B. W. (1975). *Disc. Farad. Soc.* **59**, 181.

Napper, D. and Hunter, R. J. (1975). "Hydrosols," in *Surface Chemistry and Colloids*, vol. 7, p. 161 (Ed. M. Kerker), Butterworths, London.

Ninham, B. W. and Parsegian, V. A. (1971). *J. theoret. Biol.* **31**, 405.

Ninham, B. W. and Richmond, P. (1973). *J. Chem. Soc., Farad. II* **69**, 658.

Parsegian, V. A. and Gingell, D. (1973). "Recent Advances in Adhesion" (ed. H. L. Lee). Gordon and Breach, London p. 153. [See also the review of Curtis, A. S. G. (1973) on Cell Adhesion, *Prog. in Biophys. and Molecular Biology* **27**, 315.]

Parsegian, V. A. and Weiss, G. (1972). *J. Coll. Interface Sci.* **40**, 35.

Richmond, P. (1973). *J. Phys. A* **6**, L109.

Richmond, P. and Sarkies, K. (1972). *J. Phys. C* **6**, 401.

Young, D. M. and Crowell, A. D. (1962). "Physical Adsorption of Gases." Butterworth, London.

Chapter 7

Effects of Electrolytes and Conduction Processes

7.1 Introduction

In our analysis of dispersion forces so far we have considered only such bodies and media which are non-conducting dielectrics. For a wide range of practical situations in colloid science and biology bodies interact across a medium which contains free ions. In this chapter we briefly analyse some of the effects of mobile ions or charge carriers on dispersion interactions, and will limit most of the discussion to situations where the interacting bodies are uncharged. (The corresponding problem when the surfaces of the bodies contain fixed charges or ionizable groups is clearly one of much more interest. But this area of research is as yet imperfectly explored, both theoretically and experimentally, and while some answers have been obtained see, e.g. Barnes (1975) and Barnes and Davies (1975), much work remains to be done. A useful review of work on spatially dispersive media is that of Barash and Ginzburg (1975).)

If we consider two molecules interacting across an intervening medium, as we have seen in Chapters 2 and 4, the interaction can be expressed in terms of the change in the energy of the normal modes of the field affected by the molecules. If the medium is a non-conducting dielectric the change in the field modes can be obtained from the dyadic Green function which gives the electric field due to an oscillating dipole source of frequency ω. If there are free charges present in the medium this Green function is modified in a manner which can be seen as follows.

When the source oscillates with a high frequency, the ions which are rather massive cannot follow the field, and hence would contribute little to the dielectric constant of the medium. On the other hand, at very low frequencies the motion of the ions would contribute to the dielectric constant and hence must modify the dispersion interaction. The usual method of evaluating the contributions of the ions to the dielectric constant at low frequencies is to use some variant of the linearized Poisson–Boltzmann equation (Smilga and Gorelkin, 1971; Gorelkin and Smilga, 1972a,b, 1973; Davies and Ninham, 1972). If we examine eqn. (2.79) for the free energy arising out of dispersion interaction

$$F = k_B T \sum_{n=0}^{\infty}{}' \ln D(i\xi_n); \qquad \xi_n = 2\pi n k_B T/\hbar, \qquad (7.1)$$

where $D(\omega)$ is the secular determinant whose roots give the changed normal mode frequencies of the field, it will be noticed that the $n = 0$ term corresponds to the zero frequency or static form of the secular determinant. The frequency ξ_n for $n = 1$ at room temperatures is of the order of 10^{14} rad/sec. For the usual concentrations of the electrolyte solutions of interest, say 0·1 molar, the ion plasma frequency $\omega_p = (4\pi n e^2/m)^{1/2}$ is of the order of 10^{12} rad/sec. At frequencies higher than this, the ionic plasma will have relaxed. Thus in all the terms in eqn. (7.1) for $n \neq 0$, in evaluating $D(i\xi_n)$ we can ignore the effect of ionic motion on the dielectric constants occurring therein, since even $D(i\xi_1)$ will be evaluated at a frequency considerably above the ion plasma frequency. The term $n = 0$, however, has to be dealt with separately, since in this term the dielectric constants that occur in $D(0)$ will be substantially modified by the response of the ions. This zero frequency term can be obtained from the linearized Poisson–Boltzmann equation. It should be remembered that the whole question is at issue only for interactions of organic materials across water where the similarity of uv spectra of the interacting materials makes the zero frequency contribution relatively important.

It may be stated further that a rigorous approach towards evaluating the dielectric constant of an electrolyte solution as a function of frequency is rather difficult, and requires various assumptions on the detailed dynamics of the ions (Gorelkin and Smilga, 1973; Davies and Ninham, 1972). In principle, such analyses would give corrections to the $n \neq 0$ terms in the series in eqn. (7.1). Except when geometrical constraints, to be discussed further, introduce new features, these corrections are nearly always small in practical situations, and in the main we shall confine ourselves to discussing the zero frequency term, using the linearized Poisson–Boltzmann equation.

7.2 The interaction between two molecules in an electrolyte solution

The interaction between two molecules at R_1 and R_2 in an electrolyte solution can be obtained in essentially the same way as in §2.2 or in §4.4, by evaluating the Green function $\mathscr{G}^{(M)}(\mathbf{R}_1, \mathbf{R}_1; \omega)$ defined in eqn. (4.57), with $\mathscr{G}^{(M)}$ satisfying Maxwell's equation in the dielectric medium with free charges present. In terms of this dyadic Green function, the secular determinant is, in analogy to eqn. (2.33)

$$D(\omega) = \frac{\begin{vmatrix} \mathscr{I} + 4\pi\mathscr{G}^{(M)}(\mathbf{R}_1, \mathbf{R}_1) & 4\pi\mathscr{G}^{(M)}(\mathbf{R}_1, \mathbf{R}_2) \\ 4\pi\mathscr{G}^{(M)}(\mathbf{R}_2, \mathbf{R}_1) & \mathscr{I} + 4\pi\mathscr{G}^{(M)}(\mathbf{R}_2, \mathbf{R}_2) \end{vmatrix}}{|\mathscr{I} + 4\pi\mathscr{I}^{(M)}(\mathbf{R}_1, \mathbf{R}_1)| \quad |\mathscr{I} + 4\pi\mathscr{G}^{(M)}(\mathbf{R}_2, \mathbf{R}_2)|}. \tag{7.2}$$

Using eqn. (7.1) we get

$$F = k_B T \sum_{n=0}^{\infty}{}' \ln D(i\xi_n)$$

$$= -k_B T \sum_{n=0}^{\infty}{}' (4\pi)^2 \, \mathrm{Tr}\, \{\mathscr{G}^{(M)}(\mathbf{R}_1, \mathbf{R}_2; i\xi_n)\mathscr{G}^{(M)}(\mathbf{R}_2, \mathbf{R}_1; i\xi_n)\}, \tag{7.3}$$

where we have retained terms of second order in molecular polarizability. For the reasons given in §7.1 we can put the $n \neq 0$ terms to be essentially the same as those corresponding to a non-conducting dielectric medium. The $n = 0$ term, however, will be quite different. The Green function corresponding to this term is obtained as follows. If we have a point charge source at $\mathbf{r}'$ of unit strength in the electrolyte solutions, the potential $\phi(\mathbf{r})$ satisfies Poisson's equation

$$\nabla^2 \phi(\mathbf{r}) = -\frac{4\pi}{\varepsilon(0)} \sum_{\nu} e_\nu n_\nu - \frac{4\pi}{\varepsilon} \delta(\mathbf{r} - \mathbf{r}'), \tag{7.4}$$

where $\varepsilon(0)$ is the static dielectric constant, the sum over ν is over all ionic species ν which have charge e_ν, and n_ν is the mean density of ions of species ν. This density is given by the Boltzmann relation

$$n_\nu = n_\nu(0)\mathrm{e}^{-e\phi(r)/k_B T}, \tag{7.5}$$

where $n_\nu(0)$ is the density in the absence of a perturbing potential. Expanding the exponential, using the condition for electrical neutrality $\sum_\nu e_\nu n_\nu(0) = 0$, and retaining only linear terms, eqn. (7.4) becomes

$$(\nabla^2 - \kappa_D^2)\phi(r) = -\frac{4\pi}{\varepsilon(0)} \delta(\mathbf{r} - \mathbf{r}'), \tag{7.6}$$

where κ_D is the reciprocal of the Debye screening length, given by (with n_ν standing for $n_\nu(0)$ here and subsequently)

$$\kappa_D^2 = \left(\frac{4\pi\sum_\nu n_\nu e_\nu^2}{k_B T \varepsilon(0)}\right).$$ (7.7)

The dyadic Green function corresponding to eqn. (7.6) is

$$\mathscr{G}^{(M)}(\mathbf{r}, \mathbf{r}') = -\nabla_r \nabla_r \frac{1}{(2\pi)^3 \varepsilon(0)} \int d^3k \, \frac{e^{i\mathbf{k}\cdot(\mathbf{r}-\mathbf{r}')}}{k^2 + \kappa_D^2}$$

$$= \left\{ \begin{matrix} \left(\dfrac{1}{R^2}+\dfrac{\kappa_D}{R}\right) & 0 & 0 \\[2ex] 0 & \left(\dfrac{1}{R^2}+\dfrac{\kappa_D}{R}\right) & 0 \\[2ex] 0 & 0 & -\left(\kappa_D+\dfrac{2\kappa_D}{R}+\dfrac{2}{R^2}\right) \end{matrix} \right\} \frac{e^{-\kappa_D R}}{4\pi R},$$ (7.8)

where $R = \mathbf{r} - \mathbf{r}'$. If we restrict ourselves to point molecules with polarizability $\alpha(\omega)$, at zero frequency we get from eqn. (4.57)

$$\mathscr{G}^{(M)}(\mathbf{r}, \mathbf{r}') = \alpha(0)\mathscr{G}^{(M)}(\mathbf{r}, \mathbf{r}').$$ (7.9)

Then using eqns. (7.8) and (7.9) in (7.3), the term of order $[\alpha(0)]^2$ is

$$F_0(R) \approx -\frac{\alpha^2(0)k_B T}{2\varepsilon^2(0)R^6}[2(1+\kappa_D R)^2 + (\kappa_D^2 R^2 + 2\kappa_D R + 2)^2]e^{-2\kappa_D R}$$

$$= -\frac{\alpha^2(0)k_B T}{2\varepsilon^2(0)R^6}[6 + 12\kappa_D R + 10\kappa_D^2 R^2 + 4\kappa_D^3 R^3 + \kappa_D^4 R^4]e^{-2\kappa_D R}.$$ (7.10)

If we subtract from this the result for $\kappa_D = 0$ corresponding to the absence of electrolyte, we get the change in the interaction free energy due to the electrolyte going into solution in the solvent medium,

$$\Delta F_0(r) = -\frac{k_B T\alpha^2(0)}{2\varepsilon^2(0)R^6}\{6(e^{-2\kappa_D R}-1)$$

$$+ (12\kappa_D R + 10\kappa_D^2 R^2 + 4\kappa_D^3 R^3 + \kappa_D^4 R^4)e^{-2\kappa_D R}\}.$$ (7.11)

The effect of the free ions in the electrolyte solution is thus to weaken the interaction between two small molecules due to the factor $e^{-2\kappa_D R}$ that occurs in eqn. (7.10).

7.3 Dispersion and electrostatic self-energies

The concept of the dispersion self-energy of a molecule which has been developed in Chapter 4 can easily be extended to the situation when the molecule is immersed in an electrolyte solution. Again we confine our attention to the $n = 0$ term in eqn. (7.1), and to leading order in eqn. (4.62), obtain the expression for the self-free energy of the molecule as

$$F_{SO} = (k_B T/2)[4\pi \, \text{Tr} \, \mathscr{G}^{(M)}(\mathbf{R}, \mathbf{R})]. \tag{7.12}$$

If we use a Gaussian polarization density for the molecule, as in eqn. (4.21), from eqns. (7.8) and (4.68) we get

$$
\begin{aligned}
\text{Tr} \, \mathscr{G}^{(M)}(\mathbf{R}, \mathbf{R}) &= \frac{\alpha(0)}{2(2\pi)^3} \int \frac{\text{Tr} \, (\mathbf{k}\,\mathbf{k})}{(k^2 + k_D^2)} e^{-k^2 a^2/4} \, \mathrm{d}^3 k \\
&= \frac{\alpha(0)}{4\pi^2} \left(\frac{2\sqrt{\pi}}{a^3} - \frac{2\sqrt{\pi}}{a^3}(\kappa_D a)^2 + \frac{\pi}{2a^3}(\kappa_D a)^3 \right. \\
&\qquad \left. \times \{1 - \text{erf} \, (\kappa_D a/2)\} e^{\kappa_D^2 a^2/4} \right).
\end{aligned}
\tag{7.13}
$$

If we subtract from this term the corresponding term in the absence of the electrolyte, to leading order in $\kappa_D a$ we get

$$
\begin{aligned}
\Delta F_{SO} &\approx -\frac{2 k_B T}{\sqrt{\pi} a} \kappa_D^2 \alpha(0) \\
&= -(8/\sqrt{\pi}) \frac{\alpha(0)}{a\varepsilon(0)} \left(\sum_\nu n_\nu e_\nu^2 \right).
\end{aligned}
\tag{7.14}
$$

This is a temperature–independent change in the self-free energy of a neutral molecule in an electrolyte solution. The numerical factor $(8/\sqrt{\pi})$ is not of much significance, being dependent on the nature of the assumed density distribution for polarizability.

The well-known electrostatic contribution to the free energy of an ion can be discussed within a similar mathematical framework. This energy is not due to dispersion interactions, but arises out of the electrostatic interactions of the ionic charge cloud with itself in the presence of the other ions. If $\rho_\nu(\mathbf{r})$ is a form factor which describes the charge density of an ion of the ν-th species, its electrostatic self-energy is given by

$$E_\nu = \frac{1}{2} \int \rho_\nu(\mathbf{r})\rho_\nu(\mathbf{r}') \frac{e^{-\kappa_D |\mathbf{r}-\mathbf{r}'|}}{\varepsilon(0)|\mathbf{r}-\mathbf{r}'|} \, \mathrm{d}^3 r \, \mathrm{d}^3 r'. \tag{7.15}$$

The function $[e^{-\kappa_D |\mathbf{r}-\mathbf{r}'|}/\varepsilon(0)|\mathbf{r}-\mathbf{r}'|]$ which gives the potential at $\mathbf{r}$ due to a source at $\mathbf{r}'$ is obtained from eqn. (7.6). If we assume that the charge resides on the surface of the ion whose radius is a_ν, eqn. (7.15) gives

$$E_\nu = \frac{e_\nu^2}{4\varepsilon(0)\kappa_D a^2}(1 - e^{-2\kappa_D a}) \qquad (7.16)$$

and if the electrolyte solution is dilute so that $\kappa_D a \ll 1$, we have

$$E_\nu = \frac{e_\nu^2}{2\varepsilon(0)a_\nu} - \frac{\kappa_D e_\nu^2}{2\varepsilon(0)}[1 + 0(\kappa_D a)^2]. \qquad (7.17)$$

The first term gives the Born (1920) energy of an ion immersed in a dielectric medium.* The second term gives the correction due to the electrolyte. If we sum this term over all the ions, we get

$$E = -\sum_\nu \frac{\kappa_D e_\nu^2 n_\nu}{2\varepsilon(0)}. \qquad (7.18)$$

Integration of the thermodynamic relation

$$\frac{E}{T^2} = -\frac{\partial}{\partial T}\left(\frac{F}{T}\right)$$

with the condition that as $T \to \infty$ the free energy must go to its ideal gas value gives us

$$F(\text{int}) = T\int_T^\infty \left(\frac{E}{T^2}\right) dT = -\frac{2}{3}\frac{\sqrt{\pi}}{(k_B T)^{1/2}}\left(\sum_\nu \frac{e_\nu^2 n_\nu}{\varepsilon(0)}\right)^{3/2}. \qquad (7.19)$$

This is the famous Debye–Hückel result for the free energy of interaction of a dilute electrolyte solution (Debye and Hückel, 1923; Landau and Lifshitz, 1958).

Surface energy of solutions of a strong electrolyte

The electrostatic self-energy of an ion can be used to obtain the equally well-known result of Onsager and Samaras (1934) for the change in

---

* The extension of the argument given here to include the case $\kappa_D a \approx 1$ which gives

$$E_\nu \equiv \Delta\mu_\nu = \frac{e_\nu^2 \kappa_D}{2\varepsilon(1 + \kappa_D a)}$$

can be found in the textbooks on electrochemistry. See in particular Robinson and Stokes (1950). Such expressions give good agreement with the observed activity coefficients of electrolytes, although they are obtained on the basis of the unjustified assumption that the dielectric constant $\varepsilon(0)$ retains its bulk value right up to the "surface" of the ion and of ideal solution theory. There is still no satisfactory general theory of electrolytes. In our model the activity of an ion would be

$$\Delta\mu_\nu = \frac{e^2}{4a^2\varepsilon(0)\kappa_D}(1 - 2\kappa a - e^{-2\kappa a})$$

which is practically indistinguishable from the above. The difference comes about because the usual treatment imposes the condition $\nabla^2 \phi = 0$ "inside" the ion.

surface tension of a liquid when an electrolyte is dissolved in it. We compute first the self-energy of an ion at a distance l from an interface. In Fig. 7.1 the right hand medium contains dissolved ions, and the left hand medium is assumed to be a dielectric. We require

$$E_a(l) = \frac{1}{2} \int \rho(r)\rho(r')\, d^3r\, d^3r'\phi_a(r, r'), \tag{7.20}$$

where the potential $\phi(r)$ satisfies $(z' > 0)$

$$(\nabla^2 - \kappa_D^2)\phi(\mathbf{r}, \mathbf{r}') = -\frac{4\pi}{\varepsilon_2}[\delta(\mathbf{r} - \mathbf{r}') + A_2\, \delta(z)], \qquad z > 0 \tag{7.21}$$

$$\nabla^2\phi(\mathbf{r}, \mathbf{r}') = -\frac{4\pi}{\varepsilon_1}[\delta(\mathbf{r} - \mathbf{r}') + A_1\, \delta(z)], \qquad z < 0 \tag{7.22}$$

where A_1 and A_2 are surface charge layers. We assume here that no part of the ion under consideration is allowed to spill over into the first medium. The surface charges are to be so chosen as to make it possible for ϕ to satisfy the appropriate boundary conditions on the surface. (We could include the effect of fixed surface charges also, but ignore this complication here.) The required potential can be determined by using the identity

$$\delta(\mathbf{r} - \mathbf{r}') + A_1\, \delta(z) = \frac{1}{(2\pi)^3} \int d^3k [e^{i\mathbf{k}\cdot(\mathbf{r}-\mathbf{r}')} + a_1(\kappa)e^{i\mathbf{k}\cdot\mathbf{r}}] \tag{7.23}$$

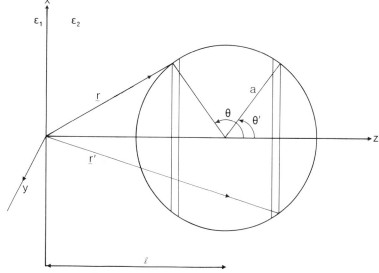

FIG. 7.1. Geometry for Onsager's problem.

with $\kappa = \sqrt{k_x^2 + k_y^2}$, whence it follows that

$$\phi_a(\mathbf{r}, \mathbf{r}') = \frac{4\pi}{(2\pi)^3 \varepsilon_2(0)} \int d^3k \, \frac{e^{i\mathbf{k}\cdot\mathbf{r}}}{(k^2 + \kappa_D^2)} [e^{-i\mathbf{k}\cdot\mathbf{r}'} + a_2(\kappa)]; \quad z > 0$$
(7.24)

$$\phi_a(\mathbf{r}, \mathbf{r}') = \frac{4\pi}{(2\pi)^3 \varepsilon_1(0)} \int d^3k \, \frac{e^{i\mathbf{k}\cdot\mathbf{r}}}{k^2} [e^{-i\mathbf{k}\cdot\mathbf{r}'} + a_1(\kappa)]; \quad z < 0. \quad (7.25)$$

Writing $k^2 + \kappa_D^2 = s^2$, and using the relation

$$2 \int_{-\infty}^{\infty} dk_z \frac{\cos k_z|z - z'|}{k_z^2 + s^2} = \frac{\pi e^{-|z-z'|s}}{s}$$
(7.26)

eqns. (7.24) and (7.25) can be rewritten as

$$\phi_a(\mathbf{r}, \mathbf{r}') = \frac{1}{2\pi\varepsilon_2} \int \frac{d^2\kappa}{s} e^{i\kappa\cdot\rho}[e^{-i\kappa\cdot\rho'}e^{-|z-z'|s} + a_2(\kappa)e^{-|z|s}]; \quad z > 0$$
(7.27)

$$\phi_a(\mathbf{r}, \mathbf{r}') = \frac{1}{2\pi\varepsilon_1} \int \frac{d^2\kappa}{\kappa} e^{i\kappa\cdot\rho}[e^{-i\kappa\cdot\rho'}e^{-|z-z'|\kappa} + a_1(\kappa)e^{-|z|\kappa}]; \quad z < 0.$$
(7.28)

The condition that ϕ and $\varepsilon(\partial\phi/\partial z)$ be continuous across the boundary gives

$$a_2(\kappa) = \left(\frac{s\varepsilon_2 - \kappa\varepsilon_1}{s\varepsilon_2 + \kappa\varepsilon_1}\right) e^{-i\kappa\cdot\rho'} e^{-|z'|s}$$
(7.29)

$$a_1(\kappa) = \frac{e^{-i\kappa\cdot\rho'}}{(s\varepsilon_2 + \kappa\varepsilon_1)} [2\varepsilon_2\kappa e^{-|z'|s} - (s\varepsilon_2 + \kappa\varepsilon_1)e^{-|z'|\kappa}]. \quad (7.30)$$

Substituting for $a_2(\kappa)$ in (7.24) we have the required potential for a source and field point both in medium 2 ($z, z' > 0$), and the self-energy of an ion of radius a is

$$E_a = \frac{1}{2} \int \rho(r)\rho(r') \frac{d^3r \, d^3r'}{2\pi\varepsilon_2(0)} \int d^2\kappa \, \frac{e^{i\kappa\cdot(\rho-\rho')}}{s}$$

$$\times \left\{ e^{-|z-z'|s} + \left(\frac{s\varepsilon_2 - \kappa\varepsilon_1}{s\varepsilon_2 + \kappa\varepsilon_1}\right) e^{-\{|z|+|z'|\}s} \right\}. \quad (7.31)$$

To evaluate the self-energy we proceed as follows. Suppose the ionic charge is distributed over the surface of the ion with density σ. We then have, in a system of spherical polar coordinates centred at the ion

$$\rho(r) \, d^3r = \sigma a^2 \sin\theta \, d\theta \, d\phi, \qquad \rho(r') \, d^3r' = \sigma a^2 \sin\theta' \, d\theta' \, d\phi'$$

$$z = (l + a\cos\theta), \qquad\qquad z' = (l + a\cos\theta') \qquad (7.32)$$

and

$$E_a = \frac{1}{2} \frac{\sigma^2 a^4}{2\pi\varepsilon_2(0)} \int \sin\theta \, d\theta \, d\phi \, \sin\theta' \, d\theta' \, d\phi' \int \frac{d^2\kappa}{s}$$

$$\times \exp\left[iak_x(\sin\theta\cos\phi - \sin\theta'\cos\phi')\right.$$

$$\left. + k_y(\sin\theta\sin\phi - \sin\theta'\sin\phi')\right]\left\{e^{-as|\cos\theta-\cos\theta'|}\right.$$

$$\left. + \left(\frac{s\varepsilon_2 - \kappa\varepsilon_1}{s\varepsilon_2 + \kappa\varepsilon_1}\right)\exp\left(-s[|l + a\cos\theta| + |l + a\cos\theta'|]\right)\right\}. \quad (7.33)$$

Using the identity

$$\int_0^{2\pi} d\phi \, e^{i(a\cos\phi + b\sin\phi)} = 2\pi J_0(\sqrt{a^2 + b^2}) \quad (7.34)$$

we get

$$E_a = \frac{(2\pi\sigma)^2 a^4}{4\pi\varepsilon_2(0)} \int_0^\pi \sin\theta \, d\theta \int_0^\pi \sin\theta' \, d\theta' \int \frac{d^2\kappa}{s} J_0(a\kappa\sin\theta)$$

$$\times J_0(a\kappa\sin\theta')\left\{e^{-|\cos\theta-\cos\theta'|sa} + \left(\frac{s\varepsilon_2 - \kappa\varepsilon_1}{s\varepsilon_2 + \kappa\varepsilon_1}\right)\right.$$

$$\left. \times e^{-s|l + a\cos\theta|}e^{-s|l + a\cos\theta'|}\right\}. \quad (7.35)$$

The first term represents the effect of the change of the potential field of the given ion due to the other ions in the presence of the surface, and is unimportant when compared with the second. The second term represents the energy associated with the image force acting on the charge a placed in a medium $\varepsilon_2(0)$ at a distance l from the surface of a medium ε_1. The image force is screened due to the presence of the other ions. To evaluate this term, we can write $|l + a\cos\theta| = l + a\cos\theta$; i.e., assume that the centre of an ion comes no closer to the surface than $l = a$, and for the surface energy have

$$E_{\text{surface}} = \sum_\nu \int_a^\infty n_\nu(l)E_\nu(l) \, dl \approx \sum_\nu n_\nu(0)\int_a^\infty E_\nu(l) \, dl, \quad (7.36)$$

where we have summed over all ionic species, and $n_\nu(0)$ is the bulk value of the density of ions far from the surface. The detailed justification for the last step is quite subtle and is given by Onsager and Samaras (1934). Since $4\pi\sigma a^2 = e_\nu$, we have after a change of variable

$$E_{\text{surface}} = \frac{e^2}{16\varepsilon_2(0)} \sum_\nu \nu^2 n_\nu(0) \int_{a\kappa_D}^\infty \frac{dy}{y} e^{-2y}\left(\int_{-1}^{+1} dx\right.$$

$$\left. \times J_0(\sqrt{y^2 - a^2\kappa_D^2}\sqrt{1-x^2})e^{-xy}\right)^2 \left(\frac{y\varepsilon_2 - \sqrt{y^2 - a^2\kappa_D^2}\,\varepsilon_1}{y\varepsilon_2 + \sqrt{y^2 - a^2\kappa_D^2}\,\varepsilon_1}\right). \quad (7.37)$$

A further approximation can be made by noting that the main contribution to the integral comes from the region $y \approx a\kappa_D$ and that for a dilute solution $a\kappa_D \ll 1$. Hence if $\varepsilon_2 \gg \varepsilon_1$, (e.g., water–air), the last factor and the Bessel function can be replaced by unity. After these approximations are made, some straightforward analysis gives

$$E_{surface} = \frac{e^2}{4\varepsilon_2(0)} \sum_\nu \nu^2 n_\nu(0) \left\{ -\ln(a_\nu\kappa_D) + \left(\frac{3}{2} - 2\ln 2 + \gamma\right) \right.$$
$$\left. + (2a\kappa_D) + \cdots \right\}, \quad (7.38)$$

where γ is Euler's constant. This tallies with the result of Onsager and Samaras. For high concentration of electrolytes or high surface charge the chief defects of the analysis is in the neglect of the variation of κ_D with distance from the surface, but the problem is still quite open and unresolved.

Surface energy of polar fluids

While the change in surface tension due to dissolved ions is given very well in terms of the above expression the actual magnitude and temperature-dependence of the surface energy of pure water or other polar fluids has never been satisfactorily calculated. In the whole of colloid science and chemical physics, it is difficult to imagine a more important problem that can be posed than the succinct deceptively simple question: What is the surface tension of water? The problem is especially frustrating, as an answer of the correct order of magnitude can be found by the following observation due to Israelachvili (1974). Recall that the dispersion contribution to surface energy can be computed by pairwise summation of the interactions of the constituent molecules, as in the method of Fowkes, or equivalently, by using Lifshitz theory at a distance of the order of an atomic spacing. The free energy so calculated gives twice the dispersion part of the surface free energy. These procedures were placed on a sound basis in Chapter 4. Thus the free energy of two semi-infinite media containing water interacting across a vacuum is given by

$$F(l) \approx -\frac{k_B T}{8\pi l^2} \sum_{n=0}^{\infty}{}' \left(\frac{\varepsilon_w(i\xi_n) - 1}{\varepsilon_w(i\xi_n) + 1}\right)^2. \quad (7.39)$$

The dispersion part is given by

$$F_{n\neq 0} \approx -\frac{\hbar}{16\pi^2 l^2} \int_0^\infty \left(\frac{\varepsilon_w - 1}{\varepsilon_w + 1}\right)^2 d\xi. \quad (7.40)$$

Taking l to be the mean interaction spacing in water we get a contribution to the surface energy of about $20\,erg/cm^2$, which is much smaller than the experimental value of $72 \, \text{erg/cm}^2$ at $0 \,^{\circ}\text{C}$. That the dispersion part is given correctly can be inferred from a number of experiments. But, as expected, Lifshitz' theory or additivity theories can not be used to give the whole story down to intermolecular distances for highly polar liquids like water where complications due to short range forces (e.g., "hydrogen bonds") are involved. At interatomic distances, the term in $n = 0$ in eqn. (7.39) gives negligible contribution when compared with the dispersion part. On the other hand, if we consider two water molecules at a distance r apart, the whole interaction energy can be expressed as

$$V(r) = V_{\text{Keesom}} + V_{\text{Debye}} + V_{\text{London}}$$

$$\approx -\frac{1}{r^6}\left(\frac{\mu^4}{3k_B T} + 2\mu^2\alpha(0) + \frac{3}{4}\alpha^2(0)\hbar\omega_{\text{uv}}\right). \tag{7.41}$$

Taking the values $\mu = 1\cdot85 \times 10^{-18}$ esu, $\alpha(0) = 1\cdot48 \times 10^{-24}$ cm^3, $\omega_{\text{uv}} = 1\cdot9 \times 10^{16}$ rad/sec, the relative ratio of the orientation energy V_{Keesom} to the dispersion energy is about 3 to 1. Hence we might expect the relative ratios of these contributions to surface-free energy to remain roughly in the same proportion and that the surface tension will decrease roughly as $1/T$ with increasing temperature. This is in fact so, but the above model is oversimplified. Unfortunately, a decomposition of the measured surface-free energy into enthalpic and entropic terms shows immediately that both the enthalpy and entropy *increases* with temperature. To construct a sound theory, it is necessary to introduce a self-consistent microscopic model for a "molecule" of water. For example, a crude model might consist of two charges e, $-e$, separated by a fixed distance a. We focus attention on one such "molecule" and replace its neighbours by a continuum with dielectric constant $\varepsilon_2(0)$. The dipole distribution will be given by some expression like

$$\rho(r) = e[f_1\delta(\mathbf{r} - \mathbf{R} - \mathbf{a}/2) - f_2\delta(\mathbf{r} - \mathbf{R} + \mathbf{a}/2)], \tag{7.42}$$

where f_1 and f_2 are atomic form factors, and $\mathbf{R}$ is the distance of the centre of the molecule from the interface. We can now calculate the self-energy of the dipole at any orientation and at a distance $\mathbf{R} = l$ from the interface. Subsequently we can carry out a statistical average over all orientations, and add up the resulting average self-energies. Unfortunately the mathematical difficulties are very great, mainly because the variation of $\varepsilon_2(0)$ as a function of distance from the interface is probably substantial in the first few layers. A self-consistent extension of e.g. the Pople model of bulk water to include ordering near a surface is badly needed; and much work remains to be done.

7.4 Effects of geometry on dispersion forces between macroscopic bodies in electrolyte solutions

Planar bodies

We turn now to the effects of electrolyte on the zero frequency term in the van der Waals force between macroscopic bodies, and consider in this section only the situation in which the surfaces of the bodies contain no fixed charges or ionizable groups. Thus the mean charge density is everywhere zero. Just as in Chapter 4, there are several equivalent approaches which can be used. In the previous section we have obtained bulk and surface-free energies by the self-energy technique, an approach which is equivalent to the standard statistical mechanical methods. The statistical mechanical method for this problem has been developed extensively by Smilga and Gorelkin (1971), Gorelkin and Smilga (1972a,b, 1973) and by Mitchell and Richmond (1973, 1974a,b). A simpler method is analogous to the van Kampen surface mode approach (Davies and Ninham, 1972).

Following the prescription given in §7.1, our starting point is the source-free Poisson-Boltzmann equation. For a homogeneous isotropic medium ε constant, and for the fluctuating potential we have

$$\nabla^2 \phi = -\frac{4\pi e}{\varepsilon(0)} \sum_\nu n_\nu e^{-e_\nu \phi / k_B T} \approx \left(\frac{4\pi}{\varepsilon(0)k_B T} \sum_\nu e_\nu^2 n_\nu\right)\phi = \kappa_D^2 \phi. \qquad (7.43)$$

In linearizing we assume that since ϕ is itself due to thermal fluctuations, the energies $e_\nu \phi$ will be less than or of the order of $k_B T$. For planar geometries, as in van Kampen's treatment for slabs, we assume solutions of the form $[\mathbf{\rho} = (x, y)]$

$$\phi(x, y, z) = f(z)e^{i\mathbf{\kappa}\cdot\mathbf{\rho}}. \qquad (7.44)$$

Then for two media "1" and "3" interacting across "2," $f(z)$ has the form

$$\begin{aligned}
f_1(z) &= A_1 e^{s_1 z}, & z < 0 \\
f_2(z) &= A_2 e^{s_2 z} + B_2 e^{-s_2 z}, & 0 < z < l \\
f_3(z) &= A_3 e^{-s_3 z}, & z > l,
\end{aligned} \qquad (7.45)$$

where the boundaries are at $z = 0$ and $z = l$, and

$$s_j^2 \equiv \kappa^2 + \kappa_D^2(j). \qquad (7.46)$$

Elimination of the coefficients by the continuity of $\phi(z)$ and $\varepsilon(\partial\phi/\partial z)$ at the interfaces gives the dispersion relation determining allowed modes as

$$D(0) = 1 - \Delta_{12}\Delta_{32} e^{-2s_2 l} = 0, \qquad (7.47)$$

where now

$$\Delta_{ij} = \frac{\varepsilon_i s_i - \varepsilon_j s_j}{\varepsilon_i s_i + \varepsilon_j s_j}. \tag{7.48}$$

The contribution to the van der Waals interaction from these modes is given by the term in $n = 0$ in (7.1), and summing over all wave vectors we have

$$F_0(l) = \frac{k_B T}{2} \int_0^\infty \frac{2\pi\kappa \, d\kappa}{(2\pi)^2} \ln \left(1 - \Delta_{12} \Delta_{32} e^{-2s_2 l}\right). \tag{7.49}$$

This differs in two ways from the zero-frequency fluctuation free energy for pure dielectrics: (1) In the replacement of the dielectric constants $\varepsilon_j = \varepsilon_j(0)$ by $\varepsilon_j s_j/\kappa = \varepsilon_j \sqrt{1 + \kappa_D^2(j)/\kappa^2}$ for medium j, and (2) in the presence of a factor s_j rather than κ wherever it occurs in an exponential. If the bodies 1 and 3 themselves have a laminar structure, the generalization of the formulae of Chapter 5 is immediate. Various asymptotic expansions for $F_0(l)$ can easily be obtained. Thus if "1" $\equiv$ "3" and these media are dielectrics, we have for $\kappa_D(2)l \gg 1$, $\varepsilon_2(0) \gg \varepsilon_1(0)$

$$F_0(l) \approx -\frac{k_B T}{16\pi l^2} \{2\kappa_D l e^{-2\kappa_D l}\} \left\{1 - \frac{4}{\varepsilon_2\sqrt{2\kappa_D l}} + \frac{1}{2\kappa_D l} + 0\left(\frac{1}{2\kappa_D l}\right)^{3/2}\right\} \tag{7.50}$$

while in the opposite limit $\kappa_D l \ll 1$, we have

$$F_0(l) \approx -\frac{k_B T}{16\pi l^2} \left(\frac{\varepsilon_1 - \varepsilon_2}{\varepsilon_1 + \varepsilon_2}\right)^2 \left\{1 - (2\kappa_D l)^2 [-\ln (2\kappa_D l) + \psi(0)]\right.$$
$$\left. \times \frac{2\varepsilon_1^2}{\varepsilon_1^2 - \varepsilon_2^2} + 0[(\kappa_D l)^3 \ln (\kappa_D l)]\right\} \tag{7.51}$$

and $\psi(t) = (d/dt) \ln \Gamma(t+1)$, $\psi(0) = 0.5722$. In the form eqn. (7.50) we see that the zero frequency attractive free energy between two planar media is that which we would have obtained if ionic conductance is neglected, but with a screening factor $2\kappa_D l e^{-2\kappa_D l}$. For biological media where $1/\kappa_D \sim 10$ Å corresponding to 0.15M salt, we can expect major modifications due to ionic conductance. In the case where the intervening medium "2" is a dielectric, and the outside media "1" and "3" contain electrolyte, there is no screening of the interaction. For the Haydon film experiment discussed in Chapter 3, this expectation is borne out.

Interactions between small bodies

We have already seen in §7.2 that the interaction of two point molecules is affected by free ions in the medium. One can expect a similar effect on the interactions between small bodies. The method of Pitaevskii (1959) can be used to extract this interaction between two small spheres in an electrolyte solution, or of two macroscopic spheres separated by a distance much larger than their radii. If the spheres themselves can contain ions (e.g. polyelectrolytes) some interesting results emerge. The method we follow is based on unpublished work of Parsegian and Ninham (1972).

Consider a dilute suspension of spheres of radius a, volume fraction v_s, whose material has dielectric constant ε_s, and characterized by a concentration parameter $n_s \equiv \sum_\nu \nu^2 n_\nu(s)$, where the sum over $n_\nu(s)$ extends over all mobile ion species contained in a sphere.* These are suspended in a medium ε_m with concentration parameter $n_m = \sum_\nu n_\nu(m)\nu^2$. Consider the interaction between two like suspensions occupying half spaces and separated by a layer of ε_m, thickness l and salt concentration n_m. We wish to treat the composite medium as an effective continuum. To do this we first write eqn. (7.43) in the form

$$\nabla.(\varepsilon\nabla\phi) = \frac{4\pi n e^2 \phi}{k_B T}; \qquad \begin{cases} n = n_s \text{ inside a sphere} \\ n = n_m \text{ inside the medium} \end{cases}. \qquad (7.52)$$

Here ε and ϕ are local quantities depending on whether ϕ is computed inside or outside the suspended particle. We can volume average both sides of this equation over regions very large compared with the intersphere distances, but much less than the wavelength variation in ϕ. This can be done, since the distance l between the two suspensions is assumed to be much greater than the sphere separation $\approx v_s^{-1/3}$. The left-hand side becomes $\varepsilon_c \nabla^2 \phi$, where c stands for composite, and (see §5.7)

$$\varepsilon_c = \varepsilon_m \left\{ 1 + 3v_s \left(\frac{\varepsilon_s - \varepsilon_m}{\varepsilon_s + 2\varepsilon_m} \right) \right\}. \qquad (7.53)$$

The right-hand side of eqn. (7.52) becomes $4\pi n_c e^2/k_B T$, with

$$n_c = n_m + v_s(n_s - n_m). \qquad (7.54)$$

Then eqn. (7.52) becomes

$$\nabla^2 \phi \equiv \kappa_D^2(c)\phi = \kappa_D^2(m) \frac{\left\{ 1 + v_s \left(\dfrac{n_s - n_m}{m_m} \right) \right\} \phi}{\left\{ 1 + 3v_s \left(\dfrac{\varepsilon_s - \varepsilon_m}{\varepsilon_s + 2\varepsilon_m} \right) \right\}}, \qquad (7.55)$$

where the inverse Debye screening length for the medium is defined by

$$\kappa_D^2(m) = \left(\frac{4\pi n_m e^2}{\varepsilon_m k_B T} \right) \equiv \kappa_m^2. \qquad (7.56)$$

* Here ν stands for the number of charge units on the ν-th ionic species.

The interaction free energy of the two composite media is

$$F_0(l) = \frac{k_B T}{4\pi} \int_0^\infty \kappa \, d\kappa \, \ln(1 - \Delta_{cm}^2 e^{-2s_m l}). \tag{7.57}$$

To find the sphere-sphere interaction, we proceed to the dilute suspension limit, i.e., $v_s \ll 1$. In that limit, the energy per unit area is expressible as a sphere-sphere interaction in the form of a pairwise sum such that

$$F_0(l) = \frac{v_s^2}{(4/3\pi a^3)^2} \int\!\!\int\!\!\int_0^\infty f_0(r) 2\pi y \, dy \, dz \, dz', \tag{7.58}$$

where $r = \sqrt{(l+z+z')^2 + x^2 + y^2}$. Differentiating both sides with respect to l we have an expression for the sphere-sphere interaction as a function of distance of separation l of the spheres which is

$$f_0(l) = -\frac{F_0'''(l)}{2\pi l} \frac{\left(\frac{4\pi}{3} a^3\right)^2}{v_s^2}. \tag{7.59}$$

Explicitly, after expanding eqn. (7.57) in powers of $v_s \ll 1$, we get

$$f_0(r) = -k_B T \frac{a^6}{r^6} e^{-2\kappa_m r} \left\{ 3E_{sm}^2 \left(1 + 2\kappa_m r + 2\kappa_m^2 r^2 + \frac{4}{3}\kappa_m^3 r^3 + \frac{2}{3}\kappa_m^4 r^4\right) \right.$$

$$\left. + E_{sm}(N_{sm} - E_{sm})\kappa_m^2 r^2 (1 + 2\kappa_m r + 2\kappa_m^2 r^2) + \frac{1}{2}(N_{sm} - E_{sm})^2 \kappa_m^4 r^4 \right\}, \tag{7.60}$$

where

$$N_{sm} = \frac{n_s - n_m}{3n_m}; \qquad E_{sm} = \frac{\varepsilon_s - \varepsilon_m}{\varepsilon_s + 2\varepsilon_m}. \tag{7.61}$$

In the absence of salt, $n_s, n_m \to 0$, $\kappa_m \to 0$ and $f_0(r)$ reduces to the usual result for two dielectric spheres

$$f_0(r) \Rightarrow -3k_B T \left(\frac{a}{r}\right)^6 \left(\frac{\varepsilon_s - \varepsilon_m}{\varepsilon_s + 2\varepsilon_m}\right)^2. \tag{7.62}$$

For high salt concentrations $\kappa_m r \to \infty$

$$f_0(r) \Rightarrow -k_B T \kappa_m^4 \frac{a^6}{2r^2} e^{-2\kappa_m r} \{N_{sm} + E_{sm}\}^2. \tag{7.63}$$

Note that the magnitude of $f_0(r)$ can be larger than what one would have expected if ion fluctuations were ignored. For the case $\varepsilon_s = \varepsilon_m$, the electrolyte free case, eqn. (7.62) predicts no attraction between the spheres, but eqn. (7.63) shows an attractive energy

$$-\frac{k_B T}{2}\left(\frac{a}{r}\right)^6 N_{sm}^2 (\kappa_m r)^4 e^{-2\kappa_m r}.$$

A similar force law emerges when the mean salt concentration is uniform, $n_s = n_m$, but $\varepsilon_s \neq \varepsilon_m$, when from (7.60)

$$f_0(r) = -3k_B T\left(\frac{a}{r}\right)^6 \left(\frac{\varepsilon_s - \varepsilon_m}{\varepsilon_s + 2\varepsilon_m}\right)^2 e^{-2\kappa_m r}$$

$$\times \left\{1 + 2\kappa_m r + \tfrac{5}{3}\kappa_m^2 r^2 + \tfrac{2}{3}\kappa_m^3 r^3 + \frac{\kappa_m^4 r^4}{6}\right\}. \quad (7.64)$$

This may be compared with eqn. (7.9) for the interaction between two molecules. Obviously the small spheres behave as molecules with polarizability $(\varepsilon_s - \varepsilon_m)/(\varepsilon_s + 2\varepsilon_m)$. The factor $e^{-2\kappa_D r}$ acts to screen the dielectric attraction eqn. (7.62). The form $r^{-2}e^{-2\kappa_m r}$ for the energy of interaction at large distances was first predicted by Kirkwood and Schumaker (1952) who were concerned with dissolved polyelectrolyte molecules at their isoelectric point. It can be considered as due to a correlation of fluctuations of the net charge on the sphere (whose time-average is zero). The suspending medium is essential, since it provides the source and sink for the charges.

Cylindrical arrays

A number of problems involving the interaction in electrolyte of anisotropic bodies, e.g. the interaction between thin rods or between a rod-like molecule and a substrate, can also be fairly easily solved by the method of Parsegian (§5.7). A variety of cases have been investigated by Parsegian and Ninham (1972) in unpublished work. The same technique has been used by Richmond (1974) to study the interaction between a rod of finite length and a substrate in the presence of salt. The analysis is straightforward, but tedious, and we shall be content with a brief resumé here. The method of derivation is traced out in Richmond's (1974) paper.

We consider long thin rods of radius a, susceptibility $\varepsilon_r^\perp$ and $\varepsilon_r^\parallel$ in the longitudinal and transverse directions, and suspended in a medium ε_m. The rods have a concentration parameter $n_r = \sum \nu^2 n_\nu(r)$ and the medium has a different concentration parameter n_m. Then for skewed rods inclined at an angle θ and at a minimum separation distance R, the zero frequency interaction free energy is

$$F_0(R, \theta) = -\frac{k_B T}{4\pi R^4} \frac{(\pi a^2)^2}{\sin \theta} \left\{ \frac{3}{4} \left(\Delta^\perp (\Delta^\perp + \Delta^\parallel) + \frac{(\Delta^\perp - \Delta^\parallel)^2}{16} \right. \right.$$

$$\times (1 + 2\cos^2 \theta) \left(1 + 2\kappa_m l + \frac{(2\kappa_m l)^2}{2!} + \frac{(2\kappa_m l)^2}{3!} \right)$$

$$+ \frac{(\kappa_m l)^2}{2} \left(N \frac{(3\Delta^\perp + \Delta^\parallel)}{2} - \frac{\Delta^\perp}{2}(\Delta^\perp + 3\Delta^\parallel) - \frac{(\Delta^\perp - \Delta^\parallel)^2}{8} \right.$$

$$\times (1 + 2\cos^2 \theta) \left((1 + 2\kappa_m l) + (\kappa_m l)^4 \left(N^2 - N(\Delta^\perp + \Delta^\parallel) \right) \right.$$

$$\left. \left. + \Delta^\perp \Delta^\parallel + \frac{(\Delta^\perp - \Delta^\parallel)^2}{8} \right) (1 + 2\cos^2 \theta) E_1(2\kappa_m l) \right\}. \qquad (7.65)$$

Here we have written $N = (n_r - n_m)/2n_m$, and $\Delta^\perp$ and $\Delta^\parallel$ are defined in eqn. (5.40). For zero ionic concentrations $N = 0$, $\kappa_m = 0$ and eqn. (7.65) reduces to the pure dielectric result (Parsegian, 1972; Mitchell and Ninham, 1973), i.e.,

$$F_0(R, \theta) = -\frac{3k_B T}{16\pi R^4} \frac{(\pi a^2)^2}{\sin \theta} \left\{ \Delta^\perp (\Delta^\perp + \Delta^\parallel) + \frac{(\Delta^\perp - \Delta^\parallel)^2}{16} (1 + 2\cos^2 \theta) \right\}. \qquad (7.66)$$

Just as for spheres, a curious result emerges when n_m, $\kappa_m \to 0$, so that all mobile charge is contained inside the rods. In that case the leading terms of $F_0(R, \theta)$ become the interaction energy for the ion-free case plus the terms

$$-\frac{k_B T}{16\pi} \frac{(\pi a^2)^2}{\sin \theta} \left\{ \left(\frac{8\pi e^2 n_r}{\varepsilon_m k_B T} \right) \frac{n_r}{R^2} (\Delta^\perp + \Delta^\parallel) + \left(\frac{8\pi e^2 n_r^2}{\varepsilon_m k_B T} \right)^2 \ln (2\kappa_m R) \right\}. \qquad (7.67)$$

This energy, due to correlation of ionic displacements along the rods, is dominated by the term $\ln (2\kappa_m R)$ as $\kappa_m \to 0$, and predicts an infinitely strong pairwise attraction at large distances of separation. It is probably of limited practical importance, chiefly because of assumptions made in the derivation which admit of no dissipation due to ionic collisions. We shall return to the possibility of such long range forces subsequently, but it is clear that forces between arrays of rods can be highly non-additive. In the opposite limit of high salt concentration in the suspending medium, the interaction behaves as

$$F(R, \theta) \approx -\frac{1}{R} e^{-2\kappa_m R}. \qquad (7.68)$$

Again this is quite different from the inverse 4th power van der Waals free energy for the ion-free case.

For parallel rods the force laws are: (1) At high ion concentration $(2\kappa_m R \gg 1)$, $F^\parallel(R)$ has the asymptotic expansion

$$F^\parallel(R) \approx -\sum_{j=0}^{\infty} \frac{b_j e^{-2\kappa_m R}}{(2\kappa_m R)^{3/2+j}} (\kappa_m a)^4 \sqrt{2\pi}(\kappa_m k_B T) \qquad (7.69)$$

where the first few coefficients are

$$b_0 = \tfrac{1}{2}(A + B + C)$$
$$b_1 = \tfrac{1}{32}(54A + 22B - 5C)$$
$$b_2 = \tfrac{1}{256}(1089A + 97B - 36C), \qquad (7.70)$$

where

$$A = \Delta^\perp(\Delta^\perp + \Delta^\parallel) + \frac{3(\Delta^\perp - \Delta^\parallel)^2}{16}$$

$$B = \frac{N(3\Delta^\perp + \Delta^\parallel)}{2} - \frac{\Delta^\perp}{2}(\Delta^\perp + 3\Delta^\parallel) - \tfrac{3}{8}(\Delta^\perp - \Delta^\parallel)^2$$

$$C = \tfrac{1}{4}(N^2 - N(\Delta^\perp + \Delta^\parallel) + \Delta^\perp\Delta^\parallel + \tfrac{3}{8}(\Delta^\perp - \Delta^\parallel)^2). \qquad (7.71)$$

On the other hand, for $\kappa_m R \to 0$, with n_r finite

$$F^\parallel(R) \Rightarrow -\frac{\pi k_B T}{32R}\left(\frac{\varepsilon_r}{\varepsilon_m}\right)^2 (\kappa_m a)^4. \qquad (7.72)$$

We remark that although the model is somewhat artificial, the very long range force predicted when $n_r \gg n_m$ suggests that in very long thin cylindrical conducting molecules, e.g., highly charged polyelectrolytes, there may very well be forces operating which are of much longer range than ordinary van der Waals forces.

7.5 Higher frequency response of an electrolyte in solution

If we were concerned solely with effects due to electrolytes on van der Waals forces, it seems a fairly safe assumption that only the very low frequency contribution will be significantly altered from predictions based on local response theory. Even at low infrared frequencies ionic mobilities are extremely small. However, proton mobilities in water are suspected to be high, and some new effects may emerge due to correlation of proton fluctuations. Conduction processes are certainly of importance in the interactions between metals, perhaps in semi-conductors, and in long chain molecules quasi-free electrons could give rise to real current fluctuations. Only fragmentary attempts have been made to construct a general theory of interactions involving materials which exhibit a non-local response function $\varepsilon = \varepsilon(k, \omega)$. Theoretical difficulties are two: (i) the absence of any knowledge of the analytic properties of $\varepsilon(k, \omega)$ in general and (ii) the proper boundary conditions for the macroscopic Maxwell equations at the boundary between a dielectric and a spatially dispersive medium are not known unambiguously, although much theoretical effort (Agarwal, Pattanayak and Wolf, 1971a,b,c: Agranovich and Ginzburg, 1966; Agranovich and Yudson, 1973; Maradudin and Mills, 1973) has gone into this problem. In the context of colloid science, the answer to (ii) is very probably that there is no answer in general. For example, if we consider the interaction of two small drops of oil (hydrocarbon) in an electrolyte, the sum of Born and dispersion self-energies for an ion inside the oil drop is so much higher than its self-energy in water that no ions can effectively enter the oil region. The proper boundary condition here can then be found from the requirement that mobile ion current at the surface must be zero. On the other hand for a protein molecule or coiled polyelectrolyte which may itself contain water, an ion can shuttle from water to the interior of the molecule with relatively much less energetic requirement, and (in an extreme case) the mobile ion current is continuous across the "surface."

Among attempts to generalize Lifshitz theory to spatially dispersive media the important papers of Craig (1972, 1973) [for similar work see Mitchell and Richmond (1973a,b)] should be mentioned. The papers of Bullough (1970) also deserve careful study.

An alternative approach is to use simple model approximations to the basic equations, in the manner indicated by Davies and Ninham (1972). If we suppose for simplicity that conduction is due to only one type of ionic species, in the non-retarded limit, Maxwell's equations can be replaced by

$$\nabla.\mathbf{E} = \frac{4\pi\rho}{\varepsilon} \tag{7.73}$$

the equation of continuity

$$\frac{\partial\rho}{\partial t} + \nabla.\mathbf{J} = 0 \tag{7.74}$$

where $\rho = ne$ is charge per unit volume, and $\mathbf{J} = ne v$ is current density; and the equation of motion for the ions in hydrodynamic approximation

$$\frac{\partial\mathbf{v}}{\partial t} + \nu\mathbf{v} + (\mathbf{v}.\nabla)\mathbf{v} = \frac{e}{m}\left(\mathbf{E} + \frac{\mathbf{v}\times\mathbf{B}}{c}\right) - \frac{\nabla p}{nm}. \tag{7.75}$$

Here p is the pressure, n the ion number density, m the ion effective mass and ν a frictional damping coefficient which models the ionic mobility. The solvent is assumed structureless with dielectric constant ε.

4/26

Carl:

You had asked to see this book.

Cost is $19.50.

We have approx. 15 days to look it over; let's say 5/10 to get it back.

Mary

Equation (7.73) is equivalent to (7.4), (7.74) is equivalent to the gauge condition on the fields, and (7.75) the simplest macroscopic approximation to the equation for vector potential. The equations can be linearized, and after Fourier analysing with respect to time, small fluctuations about equilibrium are described by the system

$$\nabla . \mathbf{E} = \frac{4\pi\rho}{\varepsilon}$$

$$\nabla . \mathbf{J} = i\omega\rho \qquad (7.76)$$

$$(-i\omega + \nu)\mathbf{J} + \frac{\omega_p^2}{\kappa_D^2}\nabla\rho = \left(\frac{\omega_p^2\varepsilon}{4\pi}\right)\mathbf{E}; \qquad \omega_p^2 = \left(\frac{4\pi n e^2}{m\varepsilon}\right); \qquad \kappa_D^2 = \left(\frac{4\pi n e^2}{k_B T}\right).$$

In an infinite homogeneous medium, these equations give the usual dielectric response functions of a dilute plasma

$$\varepsilon_l(\mathbf{k}, \omega) = \varepsilon\{1 - \omega_p^2/[\omega(\omega + i\nu) - \omega_p^2 k^2/\kappa_D^2]\}$$

$$\varepsilon_t(\mathbf{k}, \omega) = \varepsilon\{1 - \omega_p^2/[\omega(\omega + i\nu)]\}, \qquad (7.77)$$

where ε_l and ε_t are longitudinal and transverse response functions. If we consider two electrolyte solutions interacting across a dielectric "2" and impose the condition $J_z = 0$ ($z = 0, l$), then proceeding as in van Kampen's method we find (Davies and Ninham, 1972)

$$G(l, T) = \frac{k_B T}{2\pi} \sum_{n=0}^{\infty}{}' \int_0^{\infty} \kappa \, d\kappa \ln [1 - \Delta_{12}^2 \exp(-2\kappa l)], \qquad (7.78)$$

where

$$\Delta_{12} = \left(\frac{\varepsilon_1(\kappa + \gamma s) - \kappa\varepsilon_2}{(1 + \gamma)}\right) \Big/ \left(\frac{\varepsilon_1(\kappa + \gamma s) + \kappa\varepsilon_2}{(1 + \gamma)}\right)$$

$$\gamma = -\frac{\kappa\omega_p^2}{s\xi(\xi + \nu)}$$

$$s^2 = \kappa^2 + (\xi^2 + \xi\nu + \omega_p^2)\frac{\kappa_D^2}{\omega_p^2}. \qquad (7.79)$$

For small frequencies γ becomes large, $s \to \sqrt{\kappa^2 + \kappa_D^2}$, and we have

$$G_0 = \frac{k_B T}{2\pi} \int_0^{\infty} \kappa \, d\kappa \ln \left\{1 - \left(\frac{\varepsilon_1 s - \kappa\varepsilon_2}{\varepsilon_1 s + \kappa\varepsilon_2}\right) e^{-2\kappa l}\right\} \qquad (7.80)$$

in agreement with eqn. (7.49). As $\kappa_D \to 0$ we recover the result for dielectrics. On the other hand for large values of ξ, $\gamma = 0(1/\xi^3)$, $\gamma s = \kappa\omega_p^2/\xi(\xi + \nu)$. Substituting this into (7.80), the approximate dispersion relation becomes

$$\left\{1 - \left(\frac{\hat{\varepsilon} - \varepsilon_2}{\hat{\varepsilon} + \varepsilon_2}\right) e^{-2\kappa l}\right\} = 0, \tag{7.81}$$

where

$$\hat{\varepsilon} = \varepsilon\left(1 + \frac{\omega_p^2}{\xi(\xi + \nu)}\right) \tag{7.82}$$

which is equivalent to Lifshitz' prescription (Dzyaloshinskii, Lifshitz and Pitaevskii) for handling the effect of conductivity. The condition of applicability of this formula is $\nu \ll 1$. If the media "1" are dielectrics and medium "2" an electrolyte the zero frequency term is that given by (7.49). The factor corresponding to $e^{-2\kappa l}$ of the dispersion relation (7.81) at non-zero frequency is much more complicated. For further discussion see Gorelkin and Smilga (1972).

7.6 Forces of very long range?

We have already alluded to the rather surprising long range force which comes about due to low frequency fluctuations in long thin cylinders. If such cylinders or long chain molecules can be conducting in the usual sense, free or quasi-free electrons or ions can give rise to non-local susceptibilities and real current-current fluctuation correlation contributions to the free energy (as opposed to the usual polarization fluctuations). The possibility of such current correlations was suspected by London (1942) who discussed the effect of "extended harmonic oscillators." Using the π-electron model for two conjugated carbon chains, Coulson and Davies (1952) have carried out a quantum mechanical calculation of the non-retarded van der Waals forces for some specific cases, and found that the interaction was of longer range than the usual $1/R^5$ interaction for parallel chains, a conclusion which has been substantiated by other authors (Haugh and Hirshfelder, 1955; Sternlicht, 1964). Chang, Cooper, Drummond and Young (1971) have discussed the interaction between two conducting parallel chains in a vacuum at zero temperature. They found that the interaction energy per unit length behaved as $R^{-2}[\ln{(R/a)}]^{-3/2}$ where R is the distance of separation and a the radius of the chains. The corresponding expression for non-conducting chains being proportional to R^{-5}, is of much shorter range. The problem has been investigated by Davies, Ninham and Richmond (1972) using a hydrodynamic model (which ignores collisions between charge carriers). We refer to this paper for details. The results of a detailed analysis are as follows: At zero temperature, if $a \gg \lambda_D$ (where $\lambda_D = s/\omega_p$, $s = 1/m \ \partial p/\partial n$ is the isothermal sound velocity of carriers, ω_p is the plasma frequency) one has

$$E(a, R) \approx -\frac{\hbar\omega_p a}{8\sqrt{2}\pi R^2[\ln R/a]^{3/2}}.$$ (7.83)

Alternatively if $a \ll \lambda_D$

$$E(a, R) \approx -\frac{\hbar\omega_p a^4}{64\pi\lambda_D R^2}.$$ (7.84)

The low frequency contribution, which has been derived by a different method in §7.5 is of even longer range. If the chains have static susceptibilities ε_1 and the intervening medium susceptibility ε_m, we have

$$F_0(a, R) \approx -\frac{\pi k_B T}{32R}\left(\frac{\varepsilon_1}{\varepsilon_3}\right)^2\left(\frac{a}{\lambda_D}\right)^4, \qquad a \ll \lambda_D$$

$$\approx -\frac{\pi k_B T}{8R \ln^2 (R/a)}, \qquad a \gg \lambda_D.$$ (7.85)

Both results are of longer range than the zero-temperature interaction. Such results are probably very much model-dependent, and the omission of a collision mechanism is a serious defect. Fröhlich (1968, 1972) has pointed to the possibility of such very long range forces. If such contributions to interaction energies can exist the matter is clearly one of potential relevance to a variety of biological problems, e.g., viral assembly, muscle, liquid crystals, and the interaction of long thin polyelectrolytes like DNA. In all such situations it is possible that proton-proton correlations or other conducting mechanisms (π-electrons) contribute significantly. The status of such forces remains quite unclear, and we leave the question open.

7.7 The role of double layers

All of the analysis of this chapter so far has been based on the assumption that interacting surfaces are uncharged. In many situations of real interest, surfaces will bear fixed charges, or particularly in the biological domain, carry ionizable groups. As a consequence double layers will be set up and the equilibrium mobile ion distribution in an electrolyte will be inhomogeneous. The theory of electrostatic double layer interactions is well-known and such a large literature has been devoted to the subject that we shall not expand further. The chief assumptions which go into the theory are that the solvent can be replaced by a dielectric continuum, and that the Poisson–Boltzmann equation can be used. The validity of these assumptions in general is simply not known.

It has been the general practice to evaluate the electrostatic and dispersion contributions to the free energy on the assumption that they can be treated separately. The sum of these two free energies then gives a potential versus distance relation for the interaction of two bodies which is the basis of the classical theory of lyophobic colloid stability. But there is no firm basis for the separation of electrostatic and fluctuation forces which is founded in fundamental statistical mechanics, and the influence of inhomogeneity of mobile ion distribution on van der Waals forces is an important area as yet imperfectly explored. For a dilute solution of electrolyte, in view of low ionic mobility we can assert at once that only low frequency contributions to the dispersion forces will be significantly altered by the presence of charged surfaces. For each ionic species the mean ionic density will be prescribed by a Boltzmann relation

$$n_\nu(x, y, z) = n_\nu(0)e^{-e\nu\psi/k_BT}, \qquad (7.86)$$

where $n_\nu(0)$ is the averaged ionic density at a position where the potential ψ is zero. The Poisson–Boltzmann equation for the sum of the equilibrium potential ψ, and a fluctuating potential ϕ is

$$\nabla^2(\psi + \phi) = \frac{4\pi e}{\varepsilon(0)} \sum_\nu \nu n_\nu e^{-e\nu(\psi+\phi)/k_BT} \qquad (7.87)$$

whence for the fluctuating potential we have on linearizing

$$\nabla^2\phi \approx \frac{4\pi e^2}{\varepsilon(0)k_BT} \left(\sum_\nu \nu^2 n_\nu e^{-e\nu\psi/k_BT} \right) \phi. \qquad (7.88)$$

Since the equilibrium potential ψ is a complicated function of position, this equation in general is extremely difficult to solve. It has been solved exactly for planar problems, with both fixed surface charge and for charge-regulated surfaces (see Ninham and Parsegian, 1970) by Barouch, Perram and Smith (1973a,b,c,d). The latter work, a tour de force in manipulation of Weierstrassian elliptic functions is in its present form of little use for the interaction problem. They used their solutions for the normal modes and assigned an energy k_BT per mode to derive a van der Waals interaction free energy. However the direct application of van Kampen's method is not permissible. Unlike the case of uncharged surfaces, here both bulk and surface energies associated with the double layers change with separation of the bodies, due to rearrangement of the equilibrium ionic distribution approach of two charged surfaces. The problem has been attempted using a formalism similar to that of Craig (1973) by Mitchell and Richmond (1974a,b). They concluded that in the presence of double layers, the whole free energy could be computed as the sum of the usual electrostatic double layer free energy plus the van der Waals free energy for an electrolyte at uniform ionic density equal to that of the bulk electrolyte (far from any charged surfaces). However their treatment also is too restrictive, being limited to the case of extreme ionic dilution. In the language of statistical mechanics the treatments of Mitchell and Richmond, and of Barouch, Perram and Smith assume that the perturbing Hamiltonian is linear in the coupling constant, which is for this problem not correct.

The problem has been taken up by Barnes and Davies (1975) using a generalization of Craig's formalism. We refer to their work for details, and remark only that the essential difference between their work and the earlier papers mentioned is that the final results for the free energy include a coupling constant integration, in general very difficult to perform. If surface charge densities and electrolyte concentrations are not too high, for two similar flat plates, they show that the whole free energy is expressible as the sum of (i) the usual repulsive electrostatic free energy, (ii) the Onsager–Samaras surface energy contributions with corrections for excess ions in the double layer due to the presence of charges, (iii) the bulk free energy of the electrolyte, (iv) the usual screened Lifshitz result for a uniform electrolyte, and (v) an additional repulsive interaction. The theories of Barouch, Perram and Smith, and of Mitchell and Richmond emerge as different limiting cases. Of interest is the last term (v) which can be a substantial correction to the usual results.

All such attempts to understand the interaction problem require much more work, both theoretical and experimental. Further progress is probably impossible without a better understanding of the nature of water at interfaces, and of associated specific ionic effects.

References

Agarwal, G., Pattanayak, D. N. and Wolf, E. (1971a). *Phys. Rev. Letts.* **27,** 1022.
Agarwal, G., Pattanayak, D. N. and Wolf, E. (1971b). *Optics Comm.* **4,** 255.
Agarwal, G., Pattanayak, D. N. and Wolf, E. (1971c). *Optics Comm.* **4,** 260.
Agranovich, V. M. and Ginzburg, V. L. (1966). "Spatial Dispersion in Crystal Optics and the Theory of Excitons," Interscience, London.
Agranovich, V. M. and Yudson, V. I. (1973). *Optics Comm.* **7,** 121.
Barash, Yu. S. and Ginzburg, V. L. (1975). *Sov. Phys. Uspekhi* **18,** 305.
Barnes, C. (1975). Ph.D Thesis, A.N.U. Dept. of Applied Mathematics, Research School of Physical Sciences.
Barnes, C. and Davies, B. (1975). *J. Chem. Soc. Farad. Trans. II* **71,** 1667.
Barouch, E., Perram, J. W. and Smith, E. (1973a). *Chem. Phys. Letts.* **19,** 131.
Barouch, E., Perram, J. W. and Smith, E. (1973b). *Studies in Applied Mathematics* **LII,** 175.
Barouch, E., Perram, J. W. and Smith, E. (1973c). *Proc. Roy. Soc. A* **334,** 49.
Barouch, E., Perram, J. W. and Smith, E. (1973d). *Proc. Roy. Soc. A* **334,** 59.
Born, M. (1920). *Z. Physik* **1,** 45.
Bullough, R. K. (1970). *J. Phys. A* **3,** 751, and earlier papers of the same author cited therein.

Chang, D. B. Cooper, R. L., Drummond, J. E. and Young, A. C. (1971). *Phys. Letts.* **37A,** 311.

Coulson, C. A. and Davies, P. L. (1952). *Trans. Farad. Soc.* **48,** 777.

Craig, R. A. (1972). *Phys. Rev.* **B6,** 1134.

Craig, R. A. (1973). *J. Chem. Phys.* **58,** 2988.

Davies, B. and Ninham, B. W. (1972). *J. Chem. Phys.* **56,** 5797.

Davies, B., Ninham, B. W. and Richmond, P. (1973). *J. Chem. Phys.* **58,** 744.

Debye, P. and Hückel, E. (1923). *Phys. z.* **24,** 185.

Fröhlich, H. (1968). *Intern. J. Quantum Chem.* **2,** 641.

Fröhlich, H. (1972). *Phys. Letts.* **39A,** 153.

Gorelkin, V. N. and Smilga, V. P. (1972a). *Koll. Zh.* **34,** No. 10.

Gorelkin, V. N. and Smilga, V. P. (1972b). *Doklady Akad. Nauk. SSSR* **208,** 635.

Gorelkin, V. N. and Smilga, V. P. (1973). *Sov. Phys. JETP* **36,** 761 [*Z. Exp. Teor. Fiz.* **63,** 1436 (1972)].

Haugh, E. F. and Hirshfelder, J. O. (1955). *J. Chem. Phys.* **23,** 1778.

Israelachvili, J. N. (1974). *Quart. Rev. Biophys.* **6,** 341.

Kirkwood, J. G. and Shumaker, J. B. (1952). *Proc. Nat. Acad. Sci. (USA)* **38,** 863.

Landau, L. D. and Lifshitz, E. M. (1958). "Statistical Physics," Pergamon, London.

London, F. (1942). *J. Phys. Chem.* **46,** 305.

Maradudin, A. A. and Mills, D. L. (1973). *Phys. Rev.* **B7,** 2787.

Mitchell, D. J. and Ninham, B. A. (1973). *J. Chem. Phys.* **59,** 1246.

Mitchell, D. J. and Richmond, P. (1974a). *J. Coll. Interface Sci.* **46,** 118.

Mitchell, D. J. and Richmond, P. (1974b). *J. Coll. Interface Sci.* **46,** 128.

Ninham, B. W. and Parsegian, V. A. (1971). *J. theoret. Biol.* **31,** 405.

Onsager, L. and Samaras, N. T. (1934). *J. Chem. Phys.* **2,** 528.

Parsegian, V. A. (1972). *J. Chem. Phys.* **56,** 4393.

Parsegian, V. A. and Weiss, G. (1972). *J. Adhesion* **3,** 314.

Pitaevskii, L. P. (1959). *Z. Eksp. Teor. Fiz.* **37,** 577 [*Sov. Phys. JETP* **10,** 408 (1960)].

Richmond, P. (1974). *J. Chem. Soc., Farad. Trans. II* **70,** 229.

Robinson, R. A. and Stokes, R. H. (1959). "Electrolyte Solutions," 2nd ed. Butterworths, London.

Smilga, V. P. and Gorelkin, V. N. (1971). "Research in Surface Forces," v.3 (ed. B. V. Deryaguin). Consultants Bureau, N.Y. [in Russian, Nauka Press, Moscow (1967)].

Sternlicht, H. (1964). *J. Chem. Phys.* **40,** 1175.

Appendix A

A.1 Derivation of the Green function

Consider the two slab problem with media 1, 2, and 3 separated by a distance l. The z-axis is perpendicular to the slabs and the origin at the interface between media 1 and 2. The vector $\mathbf{r}'$ denotes the co-ordinate of a unit charge, and $\mathbf{r}$ denotes any field point which we take to be in any of the three media. For the moment we take $0 < z' < l$, i.e., the molecule lies in medium 2. We require solutions of the equation

$$\nabla^2 G^{(M)}(\mathbf{r}, \mathbf{r}') = \delta(\mathbf{r} - \mathbf{r}') \tag{A.1}$$

subject to the conditions that $G^{(M)}$ and $\varepsilon(\partial G^{(M)}/\partial z)$ be continuous across the boundaries. Eqn. (A.1) can be solved directly by Fourier methods, but to deal with the subsequent generalizations which include effects of retardation and/or of spatially dispersive media we prefer a slightly different technique. The normal component of the electric field $E_z \propto -(\partial G^{(M)}/\partial z)$ is discontinuous across the interfaces. Hence if $A(x, y)$, $B(x, y)$ denote the jump in E_z across the surfaces $z = 0$ and $z = l$, the electrostatic potential $G^{(M)}$ is equivalent to that due to a point charge immersed in an infinite medium 2 together with that due to surface charge densities $A(x, y)$, $B(x, y)$ at the boundaries. Recognizing this equivalence we can write

$$G^{(M)}(\mathbf{r}, \mathbf{r}') = g(\mathbf{r}, \mathbf{r}') + \int g(\mathbf{r}, \mathbf{r}'')([A(x'', y'')\,\delta(z'')$$
$$+ B(x'', y'')\,\delta(z'' - l)]\,d^3r'', \quad (A.2)$$

where

$$g(\mathbf{r}, \mathbf{r}') = -\frac{1}{4\pi\varepsilon_2|\mathbf{r} - \mathbf{r}'|}. \quad (A.3)$$

In this form the potential $G^{(M)}$ is automatically continuous across the interfaces. We now impose the remaining boundary condition that

$$\lim_{z \to (0-)} \varepsilon_1 \frac{\partial G}{\partial z} = \lim_{z \to (0+)} \varepsilon_2 \frac{\partial G}{\partial z}$$

and a similar condition at $z = l$. Use of the identity

$$\lim_{z \to (0\pm)} \frac{z}{[(x - x'')^2 + (y - y'')^2 + z^2]^{3/2}} = \pm 2\pi\,\delta(x - x'')(y - y'') \quad (A.4)$$

then gives two simultaneous integral equations for A and B which are:

$$\frac{-(\varepsilon_1 - \varepsilon_2)z'}{[(x - x')^2 + (y - y')^2 + (z')^2]^{3/2}} - 2\pi(\varepsilon_1 + \varepsilon_2)A(x, y)$$

$$= (\varepsilon_1 - \varepsilon_2)l \int_{-\infty}^{\infty} \int_{-\infty}^{\infty} \frac{dx''\,dy''B(x'', y'')}{[(x - x'')^2 + (y - y'')^2 + l^2]^{3/2}} \quad (A.5)$$

$$\frac{(\varepsilon_3 - \varepsilon_2)(l - z')}{[(x - x')^2 + (y - y')^2 + (l - z')^2]^{3/2}} + 2\pi(\varepsilon_3 + \varepsilon_2)B(x, y)$$

$$= -(\varepsilon_3 - \varepsilon_2)l \int_{-\infty}^{\infty} \int_{-\infty}^{\infty} \frac{dx''\,dy''A(x'', y'')}{[(x - x'')^2 + (y - y'')^2 + l^2]^{3/2}}. \quad (A.6)$$

These equations are readily solved by taking Fourier transforms. We define

$$A(k_1, k_2) = \frac{1}{(2\pi)^2} \int_{-\infty}^{\infty} \int_{-\infty}^{\infty} dx\,dy A(x, y)e^{-(k_1 x + k_2 y)} \quad (A.7)$$

with a similar definition for B, and have from eqns. (A.5) and (A.6),

$$A(k_1, k_2)$$
$$= \frac{1}{(2\pi)^2} \frac{(-\Delta_{12}\,\mathrm{sgn}\,z'e^{-\kappa|z'|} + \Delta_{12}\Delta_{32}\,\mathrm{sgn}\,(l - z')e^{-\kappa l}e^{-\kappa|l - z'|})}{(1 - \Delta_{12}\Delta_{32}e^{-2\kappa l})} \quad (A.8)$$

$B(k_1, k_2)$

$$= \frac{1}{(2\pi)^2} \frac{(-\Delta_{32} \operatorname{sgn} (l - z')e^{-\kappa|l-z'|} + \Delta_{12}\Delta_{32} \operatorname{sgn} z' e^{-\kappa(l+|z'|)})}{(1 - \Delta_{12}\Delta_{32}e^{-2\kappa l})} \quad (A.9)$$

where

$$\Delta_{12} = \frac{\varepsilon_1 - \varepsilon_2}{\varepsilon_1 + \varepsilon_2}, \qquad \kappa = \sqrt{(k_1^2 + k_2^2)}. \quad (A.10)$$

Hence $A(x, y)$, $B(x, y)$ are determined. Substituting the resulting expressions into eqn. (A.2) and changing to polar coordinates in k space we find with $\boldsymbol{\rho} = (x, y)$

$$G^{(M)}(\mathbf{r}, \mathbf{r}') = -\frac{1}{4\pi\varepsilon_2}\left\{\frac{1}{|\mathbf{r} - \mathbf{r}'|} + \int_0^\infty \frac{d\kappa J_0(\kappa|\boldsymbol{\rho} - \boldsymbol{\rho}'|)}{(1 - \Delta_{12}\Delta_{32}e^{-2\kappa l})} F(\kappa; z, z')\right\}, \quad (A.11)$$

where

$$F(\kappa; z, z') = \{(-\Delta_{12}e^{-\kappa(|z|+|z'|)} \operatorname{sgn} (z') + \Delta_{12}\Delta_{32} \operatorname{sgn} (l - z')$$
$$\times e^{-\kappa(|z|+|l-z'|+l)}) + [3 \rightleftharpoons 1, z \rightleftharpoons (l - z), z' \to (l - z')]\}. \quad (A.12)$$

When the position of the source, $\mathbf{r}'$, is not restricted, the corresponding generalization of eqn. (A.11) is immediate. In that case the factor $1/\varepsilon_2$ multiplying the expression in curly brackets is replaced by

$$\frac{1}{\varepsilon_2} \to \left(\frac{\theta(-z')}{\varepsilon_1} + \frac{\theta(z')\theta(l - z')}{\varepsilon_2} + \frac{\theta(z' - l)}{\varepsilon_3}\right) \equiv H(z'), \quad (A.13)$$

where $\theta(z')$ is the step function.

A.2 Evaluation of the folded Green function

We now evaluate the Green function $\tilde{G}(\mathbf{r}, \mathbf{r}')$ defined by

$$\tilde{G}(\mathbf{r}, \mathbf{r}') = \int d^3 r'' \alpha(\mathbf{r}'', \mathbf{r}') G^{(M)}(\mathbf{r}, \mathbf{r}''), \quad (A.14)$$

where

$$\alpha(\mathbf{r}'', \mathbf{r}') = \frac{\alpha(\omega)}{\pi^{3/2} a^3} e^{-(\mathbf{r}''-\mathbf{r}')^2/a^2}$$

$$= \frac{\alpha(\omega)}{(2\pi)^3} \int e^{i\mathbf{k}.(\mathbf{r}''-\mathbf{r}') - k^2 a^2/4} d^3 k. \quad (A.15)$$

Hence,

$$\tilde{G}(\mathbf{r}, \mathbf{r}') = -\frac{\alpha(\omega)}{4\pi(2\pi)^3} \int_0^\infty d\kappa J_0(\kappa|\boldsymbol{\rho} - \boldsymbol{\rho}''|) H(z'')\left(e^{-\kappa|z - z''|}\right.$$

$$\left. + \frac{F(\kappa; z, z'')}{(1 - \Delta_{12}\Delta_{32}e^{-2\kappa l})}\right) \exp\left[i\mathbf{k}' . (\mathbf{r}'' - \mathbf{r}') - k'^2 a^2/4\right] d^3 k' \, d^3 r'', \quad (A.16)$$

where we have used the identity

$$\int_0^\infty d\kappa J_0(\kappa|\boldsymbol{\rho}-\boldsymbol{\rho}''|)e^{-\kappa|z-z''|} = \frac{1}{|\mathbf{r}-\mathbf{r}''|}. \tag{A.17}$$

The integrations over x'', y'' can be carried out by writing

$$\tilde{G}(\mathbf{r},\mathbf{r}') = -\frac{\alpha(\omega)}{4\pi(2\pi)^3}\int dz'' d^3k' H(z'') \exp\left[i\mathbf{k}'\cdot(\mathbf{r}-\mathbf{r}')-k'^2a^2/4\right]$$

$$\times e^{ik_3'(z''-z)}\int d^2\rho'' e^{i\boldsymbol{\kappa}'\cdot(\boldsymbol{\rho}''-\boldsymbol{\rho})}\int_0^\infty d\kappa J_0(\kappa|\boldsymbol{\rho}-\boldsymbol{\rho}''|)$$

$$\times\left(e^{-\kappa|z-z''|}+\frac{F(\kappa;z,z'')}{(1-\Delta_{12}\Delta_{32}e^{-2\kappa l})}\right)$$

$$= -\frac{\alpha(\omega)}{4\pi(2\pi)^2}\int_{-\infty}^\infty dz'' \int d^3k' H(z'')$$

$$\times\exp\left\{i[\mathbf{k}'\cdot(\mathbf{r}-\mathbf{r}')-k'^2a^2/4+ik_3'(z''-z)]\right\}\int_0^\infty dk\left(e^{-\kappa|z-z''|}\right.$$

$$\left.+\frac{F(\kappa;z,z'')}{(1-\Delta_{12}\Delta_{32}e^{-2\kappa l})}\right)\int_0^\infty \rho\,d\rho J_0(\kappa'\rho)J_0(\kappa\rho). \tag{A.18}$$

It is straightforward to show that

$$\int_0^\infty \rho\,d\rho J_0(\kappa'\rho)J_0(\kappa\rho) = \frac{\delta(\kappa'-\kappa)}{\kappa} \tag{A.19}$$

so that $\tilde{G}(\mathbf{r},\mathbf{r}')$ becomes

$$\tilde{G}(\mathbf{r},\mathbf{r}') = -\frac{\alpha(\omega)}{4\pi(2\pi)^2}\int_{-\infty}^\infty dz''\int d^3k H(z'')$$

$$\times\exp\left[i\mathbf{k}\cdot(\mathbf{r}-\mathbf{r}')-k^2a^2/4+ik_3(z''-z)\right]\frac{1}{\kappa}\left(e^{-\kappa|z-z''|}\right.$$

$$\left.+\frac{F(\kappa;z,z'')}{(1-\Delta_{12}\Delta_{32}e^{-2\kappa l})}\right). \tag{A.20}$$

The integration over k_3 is again immediate, and we find

$$\tilde{G}(\mathbf{r},\mathbf{r}') = -\frac{\alpha(\omega)}{8\pi^{5/2}}\int_{-\infty}^\infty dz'' H(z'')\int d^2\kappa\,\exp\left[i\boldsymbol{\kappa}\cdot(\boldsymbol{\rho}-\boldsymbol{\rho}')-\kappa^2a^2/4\right]$$

$$\times\frac{1}{\kappa}\left(e^{-\kappa|z-z''|}+\frac{F(\kappa;z,z'')}{(1-\Delta_{12}\Delta_{32}e^{-2\kappa l})}\right)e^{-(z''-z')^2/a^2}. \tag{A.21}$$

To evaluate the force between media and energies of interaction we need now the trace of the double gradient of $\tilde{G}$ evaluated at $\mathbf{r} = \mathbf{r}'$. This follows directly from eqn. (A.20), and we have

$$\mathrm{Tr}\,\mathscr{G}^{(M)}(\mathbf{r}, \mathbf{r}; \omega) = \mathrm{Tr}\,\nabla_r \nabla_{r'} \tilde{G}(\mathbf{r}, \mathbf{r}')\big|_{r' \to r} = \frac{\alpha(\omega)}{16\pi^3} \int d^3k$$

$$\times e^{-k^2/a^2/4} e^{-ik_3 z}\left\{ \kappa^2 I(\kappa, z) - \frac{\partial^2}{\partial z^2} I(\kappa, z) \right\}, \quad (A.22)$$

where

$$I(\kappa, z) = \int_{-\infty}^{\infty} dz'' H(z'') e^{ik_3 z''} \frac{1}{\kappa}\left(e^{-\kappa|z-z''|} + \frac{F(\kappa; z, z'')}{(1 - \Delta_{12}\Delta_{32} e^{-2\kappa l})} \right). \quad (A.23)$$

After carrying out the k_3 integrations and then those over z'', we find, after some tedious algebra, the result quoted in eqn. (4.106).

Appendix B

In this appendix we shall derive a more exact form of the interaction Hamiltonian of eqn. (4.4) in the non-retarded regime.

The relationship between an arbitrary electric field $\mathbf{E}(\mathbf{r}, \omega)$ and the corresponding potential $\phi(\mathbf{r}, \omega)$ is

$$\mathbf{E}(\mathbf{r}, \omega) = -\nabla \phi(\mathbf{r}, \omega). \tag{B.1}$$

Expressing both sides as Fourier integrals we get

$$\frac{1}{(2\pi)^3} \int e^{i\mathbf{k}\cdot\mathbf{r}} \mathbf{E}(\mathbf{k}, \omega) \, d^3k = -\nabla \frac{1}{(2\pi)^3} \int e^{i\mathbf{k}\cdot\mathbf{r}} \phi(\mathbf{k}, \omega) \, d^3k$$

$$= -\frac{1}{(2\pi)^3} \int (ik) \phi(\mathbf{k}, \omega) e^{i\mathbf{k}\cdot\mathbf{r}} \, d^3k. \tag{B.2}$$

Hence

$$\mathbf{E}(\mathbf{k}, \omega) = -i\mathbf{k}\phi(\mathbf{k}, \omega),$$

or

$$\phi(\mathbf{k}, \omega) = \frac{i\mathbf{k}\cdot\mathbf{E}(\mathbf{k}, \omega)}{k^2}. \tag{B.3}$$

Then,

$$\phi(\mathbf{r}, \omega) = \frac{1}{(2\pi)^3} \int e^{i\mathbf{k}\cdot\mathbf{r}} \phi(\mathbf{k}, \omega) \, d^3k$$

$$= \frac{i}{(2\pi)^3} \int e^{i\mathbf{k}\cdot\mathbf{r}} \frac{\mathbf{k}\cdot\mathbf{E}(\mathbf{k}, \omega)}{k^2} \, d^3k. \tag{B.4}$$

For a dipole with charge $(+e)$ at $\mathbf{R}$, and $(-e)$ at $\mathbf{R}+\mathbf{u}$,

$$H_{\text{int}} = e\phi(\mathbf{R}, \omega) - e\phi(\mathbf{R}+\mathbf{u}, \omega)$$

$$= \frac{ei}{(2\pi)^3} \int \frac{\mathbf{k}.\mathbf{E}(\mathbf{k}, \omega)}{k^2} e^{i\mathbf{k}.\mathbf{R}} [1 - e^{i\mathbf{k}.\mathbf{u}}] \, d^3k. \tag{B.5}$$

If the electric field has the form given in eqn. (4.1) only one k-component will occur, and then eqn. (B.5) reduces to,

$$H_{\text{int}} = ei \frac{\mathbf{k}.\mathbf{E}(\mathbf{k}, \omega)}{k^2} e^{i\mathbf{k}.\mathbf{R}} [1 - e^{i\mathbf{k}.\mathbf{u}}]. \tag{B.6}$$

With this interaction Hamiltonian, which differs from that given in eqn. (4.4), the form of $\boldsymbol{\alpha}_n(\mathbf{k}, \omega)$ of eqn. (4.8) becomes

$$\boldsymbol{\alpha}_n(\mathbf{k}, \omega) \equiv -\frac{e^2}{\hbar} \sum_m \left(\frac{\langle n| \mathbf{u} |m\rangle\langle m| e^{i\mathbf{k}.\mathbf{u}} |n\rangle}{\omega_{nm} + \omega} \right.$$
$$\left. + \frac{\langle m| \mathbf{u} |n\rangle\langle n| e^{i\mathbf{k}.\mathbf{u}} |m\rangle}{\omega_{nm} - \omega} \right) \frac{i\mathbf{k}}{k^2}. \tag{B.7}$$

The dyadic terms here are $\langle n| \mathbf{u} |m\rangle\mathbf{k}$ and $\langle m| \mathbf{u} |n\rangle\mathbf{k}$. Although this form of $\boldsymbol{\alpha}_n(\mathbf{k}, \omega)$ differs from that given in eqn. (4.8), the subsequent arguments are unaffected if a phenomenological form for $\boldsymbol{\alpha}(\mathbf{r}, \omega)$, such as in eqn. (4.20), is used.

Appendix C

This appendix is concerned with the derivation of eqn. (4.54).

If we consider a point dipole source of strength $\mathbf{p}$ at $\mathbf{r}'$, the field due to that at $\mathbf{r}$ is given by

$$\mathbf{E}(\mathbf{r}) = -4\pi\mathcal{G}(\mathbf{r}, \mathbf{r}')\mathbf{p}. \tag{C.1}$$

If there are other molecules present, they will be polarized, and the field at $\mathbf{r}$ will then be the sum of the field obtained above, and the fields due to the polarized molecules. The latter will be obtained as follows. The polarization density of the l-th molecule is

$$\mathbf{p}_l(\mathbf{r}) = \alpha(\mathbf{r} - \mathbf{R}_l)\mathbf{E}(\mathbf{R}_l). \tag{C.2}$$

The field due to this molecule is,

$$-4\pi \int \mathcal{G}(\mathbf{r}, \mathbf{r}')\alpha(\mathbf{r}' - \mathbf{R}_l)\mathbf{E}(\mathbf{R}_l) \, \mathrm{d}^3 r' = -4\pi\mathcal{G}(\mathbf{r}, \mathbf{R}_l)\mathbf{E}(\mathbf{R}_l). \tag{C.3}$$

Thus we get a set of self-consistent equations for the fields,

$$\mathbf{E}(\mathbf{r}) = -4\pi\mathcal{G}(\mathbf{r}, \mathbf{r}')\mathbf{p} - 4\pi \sum_l \mathcal{G}(\mathbf{r}, \mathbf{R}_l)\mathbf{E}(\mathbf{R}_l). \tag{C.4}$$

$\mathbf{E}(\mathbf{R}_l)$ can obviously be solved in the form

$$\mathbf{E}(\mathbf{R}_l) = -\sum_j [\mathcal{I} + 4\pi\mathcal{G}]_{lj}^{-1} 4\pi\mathcal{G}(\mathbf{R}_j, \mathbf{r}')\mathbf{p}. \tag{C.5}$$

If we multiply both sides of this equation by $\boldsymbol{\alpha}(\mathbf{r}-\mathbf{R}_l)$ and integrate over $\mathbf{r}$, using eqn. (4.9), we get

$$\mathbf{P}_l = -\left[\int \boldsymbol{\alpha}(\mathbf{r}-\mathbf{R}_l)\,\mathrm{d}^3r\right]\sum_j [\mathscr{I}+4\pi\mathscr{G}]^{-1}_{lj}\,4\pi\mathscr{G}(\mathbf{R}_j,\mathbf{r}')\mathbf{p}. \qquad (C.6)$$

Now, if the external source $\mathbf{p}$ at $\mathbf{r}'$ has the same polarizability density as the other molecules, we get

$$\mathscr{G}(\mathbf{R}_j,\mathbf{r}')\mathbf{p} \Rightarrow \int \mathscr{G}(\mathbf{R}_j,\mathbf{r}'')\boldsymbol{\alpha}(\mathbf{r}''-\mathbf{r}')\mathbf{E}(\mathbf{r}')\,\mathrm{d}^3r''$$

$$= \mathscr{G}(\mathbf{R}_j,\mathbf{r}')\mathbf{E}(\mathbf{r}'). \qquad (C.7)$$

The integral $\int\boldsymbol{\alpha}(\mathbf{r}-\mathbf{R}_l)\,\mathrm{d}^3r$ is independent of l and gives the total polarizability of any one of the molecules. If this total polarizability tensor commutes with the dyadic Green function (which is assumed to be the case), and if eqn. (C.7) is used in (C.6), we get

$$\mathbf{P}_l = -\sum_j [\mathscr{I}+4\pi\mathscr{G}]^{-1}_{lj}\,4\pi\mathscr{G}(\mathbf{R}_j,\mathbf{r}')\left[\int \boldsymbol{\alpha}(\mathbf{r}-\mathbf{r}')\,\mathrm{d}^3r\right]\mathbf{E}(\mathbf{r}')$$

$$= -\sum_j [\mathscr{I}+4\pi\mathscr{G}]^{-1}_{lj}\,4\pi\mathscr{G}(\mathbf{R}_j,\mathbf{r}')\mathbf{P}_{\text{ext}}. \qquad (C.8)$$

This is eqn. (4.54), if $\mathbf{r}'$ is set equal to $\mathbf{R}_l$, the position of the external source.

C.1 Extension of §4.7 to Dense Systems

For simplicity in Chapter 4, we have $\varepsilon = 1+4\pi n\alpha$, where α is the vacuum polarizability of a molecule. This is true only for a dilute system. In a dense system the local field experienced by an embedded molecule is not given through the macroscopic Green function $\mathscr{G}^{(M)}$, but by the sum of the field due to $\mathscr{G}^{(M)}$ and that due to the induced (surface) charges on the cavity surrounding the molecule. For example for a spherical cavity the total field is obtained through an effective macroscopic Green function $\mathscr{G}^{(M)}_{\text{dense}} = (\varepsilon+2)/3\,\mathscr{G}^{(M)}$. If the additional factor is retained throughout the analysis in §4.7 and if in (4.92) the Clausius-Mossotti formula for $\alpha_j(\omega)$, i.e. $\alpha_j(\omega) = \dfrac{3}{4\pi}\left(\dfrac{\varepsilon-1}{\varepsilon+1}\right)$ is used, then eqns. (4.94) and hence eqn. (4.99) will be unchanged. The argument holds whatever be the shape of the cavity surrounding the molecule so that eqn. (4.99) is of general validity.

Appendix D

Consider the sum which occurs in eqns. (6.22)

$$S_1 = \sum_{j=1}^{\infty} e^{-Bj - d/j^2},$$
(D.1)

where we write $p/p_0 \equiv e^{-B} \equiv z$, and deal first with the case $B = \ln(p_0/p) \ll 1$. When further $d \gg 1$, the sum can be replaced by an integral if and only if $(B^2 d)^{1/3} \gg 1$. Evaluating this integral by the method of steepest descents we find

$$S_1 = \left(\frac{2^{4/3}\pi}{3}\right)^{1/2} \left(\frac{d}{B}\right)^{1/3} \frac{1}{(B^2 d)^{1/6}} e^{-(3/2)(2B^2 d)1/3}$$

$$\times \left\{1 + 0\left(\frac{1}{(B^2 d)^{1/3}}\right)\right\}, \qquad B \ll 1, \qquad d \gg 1, \qquad (B^2 d)^{1/3} \gg 1. \quad (D.2)$$

On the other hand, when the last inequality is reversed we write

$$S_1 = \sum_{j=1}^{\infty} e^{-Bj} \sum_{m=0}^{\infty} \frac{(-1)^m d^m}{m! \, j^{2m}}$$

$$= \frac{e^{-B}}{1 - e^{-B}} + \sum_{m=1}^{\infty} \frac{(-1)^m d^m}{m!} e^{-B} \Phi(e^{-B}, 2m, 1), \quad (D.3)$$

where the function $\Phi(z, s, v)$ is defined by (Erdelyli *et al.*, 1953)

$$\Phi(z, s, v) = \sum_{j=1}^{\infty} \frac{z^j}{(j+v)^s}, \qquad |z| < 1. \tag{D.4}$$

This function has the convergent expansion

$$z\Phi(z, n, 1) = \sum_{k=0}^{\infty}{}' \zeta(n-k) \frac{(\ln z)^k}{n!} + \frac{(\ln z)^{n-1}}{(k-1)!}$$
$$\times [\psi(n) - \psi(1) - \ln(\ln 1/z)], \qquad |\ln z| < 2. \tag{D.5}$$

Here the prime indicates that the term in $k = n-1$ is to be omitted, $\zeta(n-k)$ is the Riemann zeta function, and $\psi(z) = d/dz \ln \Gamma(z)$ is the logarithmic derivative of the gamma function. Under the conditions $B \ll 1, d \gg 1, (B^2 d) \ll 1$ using eqn. (D.5) in (D.3) we have

$$S_1 \approx \frac{1}{B} + \sum_{m=1}^{\infty} \frac{(-1)^m d^m}{m!} \zeta(2m)[1 + 0(B)]$$
$$- \frac{1}{B} \sum_{m=1}^{\infty} \frac{(-1)^m (dB^2)^m}{m!(2m-1)!} [\psi(m) - \psi(1) - \ln B]. \tag{D.6}$$

The second sum can be approximated for large d as

$$\sum_{m=1}^{\infty} \frac{(-1)^m d^m}{m!} \zeta(2m) = \sum_{m=1}^{\infty} (e^{-d/m^2} - 1)$$
$$\approx \int_0^{\infty} (e^{-d/m^2} - 1) \, dm \approx -\frac{\sqrt{\pi d}}{2}. \tag{D.7}$$

Hence

$$S_1 \approx \frac{1}{B} \left\{ 1 - \frac{\sqrt{\pi}}{2}(B^2 d)^{1/3} - (B^2 d) \ln B + 0(B^2 d) \right\},$$
$$B \ll 1, \qquad d \gg 1, \qquad B^2 d \ll 1. \tag{D.8}$$

Corresponding to eqn. (D.2) and (D.8) we require also expressions for

$$S_2 = \sum_{j=1}^{\infty} j e^{-Bj - d/j^2}.$$

We find in the same way that

$$S_2 \approx (d/B)^{1/3} \sigma_1, \qquad B \ll 1, \qquad d \gg 1, \qquad (B^2 d) \gg 1 \tag{D.9}$$

$$S_2 \approx \frac{1}{B} \{ 1 + \tfrac{1}{2}(B^2 d) \ln(B^2 d) + \cdots \} \tag{D.10}$$

for $B \ll 1, d \gg 1, B^2 d \ll 1$.

Postscript

Since the completion of this manuscript in early 1974, a number of developments have taken place in the general area of dispersion forces. To indicate some trends in the research effort in this field we shall make a few remarks and give a few references which are by no means exhaustive.

General theory

The semi-classical approach we have followed in this book is adequate to describe the main features of dispersion forces between molecules. But in the presence of dielectrics and conductors a rigorous formulation must be based on quantum electrodynamics. Such a formulation using suitable response functions of the electromagnetic field has been given by Agarwal (1975a,b,c).

A substantial amount of recent effort has been directed towards understanding those properties of metals which determine the dispersion force between a molecule and a metal, the force between two metallic bodies, or the force between adsorbate molecules on a metal surface. The important property of a metal here is its dielectric response. The problem of evaluating the dielectric susceptibility of a metal taking into consideration boundary effects and spatial dispersion (i.e., the wave number dependence of the dielectric constant) has received considerable attention lately. The properties of an inhomogeneous electron gas (such as would obtain in the surface region of a semi-infinite metal block) has been studied by Mukhopadhyay and Lundqvist (1975a). Heinrichs (1975) has studied the Lifshitz force between two metal slabs using special boundary conditions on the motion of electrons at the surfaces.

In non-conducting dielectric bodies the major part of the surface energy is from dispersion forces. In the case of metals, however, the contribution of the metallic electrons to the surface energy is rather involved. Various aspects of this contribution and their relevance to dispersion interactions between metals have been studied lately. We mention in particular the work of Inglesfield and Wikborg (1975), Jonson and Srinivasan (1974), Harris and Jones (1974), Mahan (1975), Langbein (1975a,b) and Chan and Richmond (1976).

The interaction of a molecule with a metal surface taking into consideration the properties of the semi-infinite electron gas representing the metal has been studied by Zaremba and Kohn (1976). The dispersion interaction between voids in the metal has been studied, among others, by Mukhopadhyay and Lundqvist (1975b). A recent most useful review is that of Barash and Ginzburg (1975). The problem of interaction of and between molecules on metals is likely to remain a growth area because of its importance to understanding catalysis.

The experimental situation

Not many new experiments have been performed on the direct measurement of dispersion forces. Experiments dealing with measurement of the force between alkali atoms and noble metal surfaces have been reported by Shih (1974) and Shih and Parsegian (1975). In the latter experiment a 60% discrepancy between the predictions of the general Lifshitz theory and the observed force is still not understood. The first of a most interesting series of indirect experiments have been reported by LeNeveu, Rand and Parsegian (1976a) and LeNeveu, Rand, Gingell and Parsegian (1976b), who worked with the multilayer lecithin-water system. The repulsive forces here are not electrostatic in the usual sense in that they are due to water structure and have been accounted for in a phenomenological theory by Marčelja and Radić (1976). The attractive forces which remain when due allowance is made for these repulsive forces are in agreement with theoretical estimates for van der Waals forces in multilayers discussed in Chapter 5. While short-range [0 (10 Å)] water structure forces are beyond our brief, it should be mentioned that recent ideas taken from advances in our theoretical understanding of liquid crystals are likely to prove extremely fruitful in developing at least simple phenomenological theories in this area. A recent extension of the Marčelja mean field model of water structuring forces explains the near ideality of activity of aqueous solutions of hydrophilic molecules (Forsyth, Marčelja, Mitchell and Ninham, 1976).

The status of the extensive studies of Haydon (1976) and his collaborators on oil-water systems has recently been reported. Parallel studies by P. M. Krugylakov and his collaborators from the Novosibirsk School will also be available shortly.

The experiments of Israelachvili and Tabor on van der Waals forces between mica surfaces across vacuum have been subjected to a critical theoretical re-evaluation (White, Israelachvili and Ninham, 1976; Chan and Richmond, 1976). The latter used very detailed far-ultraviolet data which remove theoretical uncertainties and showed that there are discrepancies between theory and experiment. The former authors traced these discrepancies to a systematic experimental error in the least expected place. The upshot of these analyses is that when allowance is made for this error, and for the presence of one or two adsorbed water layers, theory and experiment are in the end in very good agreement indeed. It is clear from this work that the older notion of a Hamaker "constant" is invalid, although still a useful guide to thinking.

The experimental situation with regard to the status of dispersion forces across condensed media up to 1975 has been reviewed by Napper and Hunter (1975). This very difficult experimental problem has been tackled by Israelachvili (1976a,b). He has succeeded in measuring directly (with a distance resolution of 1–2 Å) the forces between mica surfaces across aqueous solutions of KNO_3 of various concentrations ($10^{-4} - 1$ Molar). Although still in a preliminary stage, it is already apparent that down to 20 Å separation at least, the repulsive electrostatic forces are given to considerable accuracy by the Poisson-Boltzmann equation. Likewise the attractive forces are given accurately by the general theory of van der Waals forces. Developments in this area are awaited with much interest.

As a final comment, we remark that the role of dispersion forces in colloid science has been reviewed by Israelachvili and Ninham (1976). We list last some other reviews which deal with dispersion forces from various emphases; That of Dzyaloshinskii, Lifshitz and Pitaevskii is of course seminal and unlikely to be superseded.

Other reviews on the theory of van der Waals forces

Dzyaloshinski, I. E., Lifshitz, E. M. and Pitaevskii, L. P. (1961). *Advan. Phys.* **10,** 165.

Barash, Yu. S. and Ginzburg, V. L. (1975). *Sov. Phys. Uspehki* **18,** 305.

Israelachvili, J. N. (1974). *Quart. Rev. Biophys.* **6,** 341.

Krupp, H. (1967). *Advan. Coll. Interface Sci.* **1,** 111.

Langbein, D. (1974). "Theory of van der Waals Attraction," Springer Tracts in Modern Physics, Springer-Verlag, Berlin, vol. 72.

Richmond, P. (1975), in Specialist Periodical Reports, "Colloid Science," vol. 2 (ed. D. H. Everett), Chemical Society.

Parsegian, V. A. (1973). *Ann. Rev. Biophys. Bioeng.* 2, 221.

References

Agarwal, G. S. (1975a). *Phys. Rev.* A**11,** 230.

Agarwal, G. S. (1975b). *Phys. Rev.* A**11,** 243.

Agarwal, G. S. (1975c). *Phys. Rev.* A**11,** 253.

Barash, Yu. S. and Ginzburg, V. L. (1975). *Sov. Phys. Uspekhi* **18,** 305.

Chan, D. and Richmond, P. (1975a). *J. Phys. C* **8,** 2509.

Chan, D. and Richmond, P. (1975b). *J. Phys. C* **8,** 3221.

Chan, D. and Richmond, P. (1976a). *J. Phys. C* **9,** 153.

Chan, D. and Richmond, P. (1976b). *J. Phys. C* **9,** 163.

Chan, D. and Richmond, P. (1976c). Preprint, Unilever Research Laboratories, to appear in *Proc. Roy. Soc.*

Forsyth, P., Jr., Marčelja, S., D. J. Mitchell and Ninham, B. W. (1976). *Chem. Phys. Letts.* (to appear).

Harris, J. and Jones, R. O. (1974). *J. Phys. F* **4,** 1170.

Haydon, D. A. and collaborators (1975).

See J. Requena, D. F. Billett and D. A. Haydon (1975). *Proc. Roy. Soc.* A**347,** 141.

J. Requena and D. A. Haydon (1975). *Proc. Roy. Soc.* A**347,** 161.

D. E. Brooks, Y. K. Levine, J. Requena and D. A. Haydon (1975). *Proc. Roy. Soc. Lond.* A**347,** 179.

Heinrichs, J. (1975). *Phys. Rev.* B**11,** 3625; B**11,** 3637.

Inglesfield, J. E. and Wikborg, E. (1975). *J. Phys. F* **5,** 1475.

Israelachvili, J. N. and Adams, G. E. (1976). *Nature* **262,** 774.

Israelachvili, J. N. and Ninham, B. W. (1976), in Proceedings of the 50th Anniversary Meeting of ACS Colloid and Surface Science Division, lecture entitled "Intermolecular Forces—The Long and Short of it," to appear in *J. Coll. Interface Sci.*

Jonson, M. and Srinivasan, G. (1975). *Physica Scripta* **10,** 262.

Langbein, D. (1975). *J. Phys. A* **8,** 1593.

Langbein, D. (1976). *J. Phys. A* **9,** 627.

LeNeveu, D. M., Rand, R. P., Gingell, D. and Parsegian, V. A. (1975). *Science* **191,** 399.

LeNeveu, D. M., Rand, R. P. and Parsegian, V. A. (1976). *Nature* **259,** 601.

Mahan, G. D. (1975). *Phys. Rev.* B**12,** 5585.

Marčelja, S. and Radić, N. (1976). *Chem. Phys. Letts.* to appear.

Mukhopadhyay, G. and Lundqvist, S. (1975a). *Nuovo Cim.* **27**B 1.

Mukhopadhyay, G. and Lundqvist, S. (1975b). *Solid State. Comm.* **17,** 949.

[For some other aspects of surface phenomena, see Surface Science, Parts I and II (lectures presented at an International course at ICTP, Trieste, 16 Jan.–10 April (1974), Vienna, Austria, IAEA (1975)).]

Napper, D. and Hunter, R. J. (1975). "Hydrosols," in Surface Chemistry and Colloids, vol. 7 (M. Kerker, ed.), Butterworths, London.

Shih, A. (1975). *Phys. Rev.* A**9,** 1507.

Shih, A. and Parsegian, V. A. (1975). *Phys. Rev.* A**12,** 835.

White, L. R., Israelachvili, J. N. and Ninham, B. W. (1976). Preprint, Department of Applied Mathematics, ANU, to appear in *J. Chem. Soc., Faraday I*

Zaremba, E. and Kohn, W. (1976). *Phys. Rev.* **B**13, 2270.

Subject Index